THE GOAL SHOULD BE, NOT TO IMPLANT IN THE STUDENTS' MIND EVERY FACT THAT THE TEACHER KNOWS NOW; BUT RATHER TO IMPLANT A WAY OF THINKING THAT ENABLES THE STUDENT, IN THE FUTURE, TO LEARN IN ONE YEAR WHAT THE TEACHER LEARNED IN TWO YEARS. ONLY IN THAT WAY CAN WE CONTINUE TO ADVANCE FROM ONE GENERATION TO THE NEXT.

EDWIN JAYNES

THE IMPORTANT THING IN SCIENCE IS NOT SO MUCH TO OBTAIN NEW FACTS AS TO DISCOVER NEW WAYS OF THINKING ABOUT THEM.

SIR WILLIAM LAWRENCE BRAGG

JAKOB SCHWICHTENBERG

PHYSICS FROM FINANCE

NO-NONSENSE BOOKS

no-nonsense books

First printing, May 2020

BOOK EDITION: 1.8

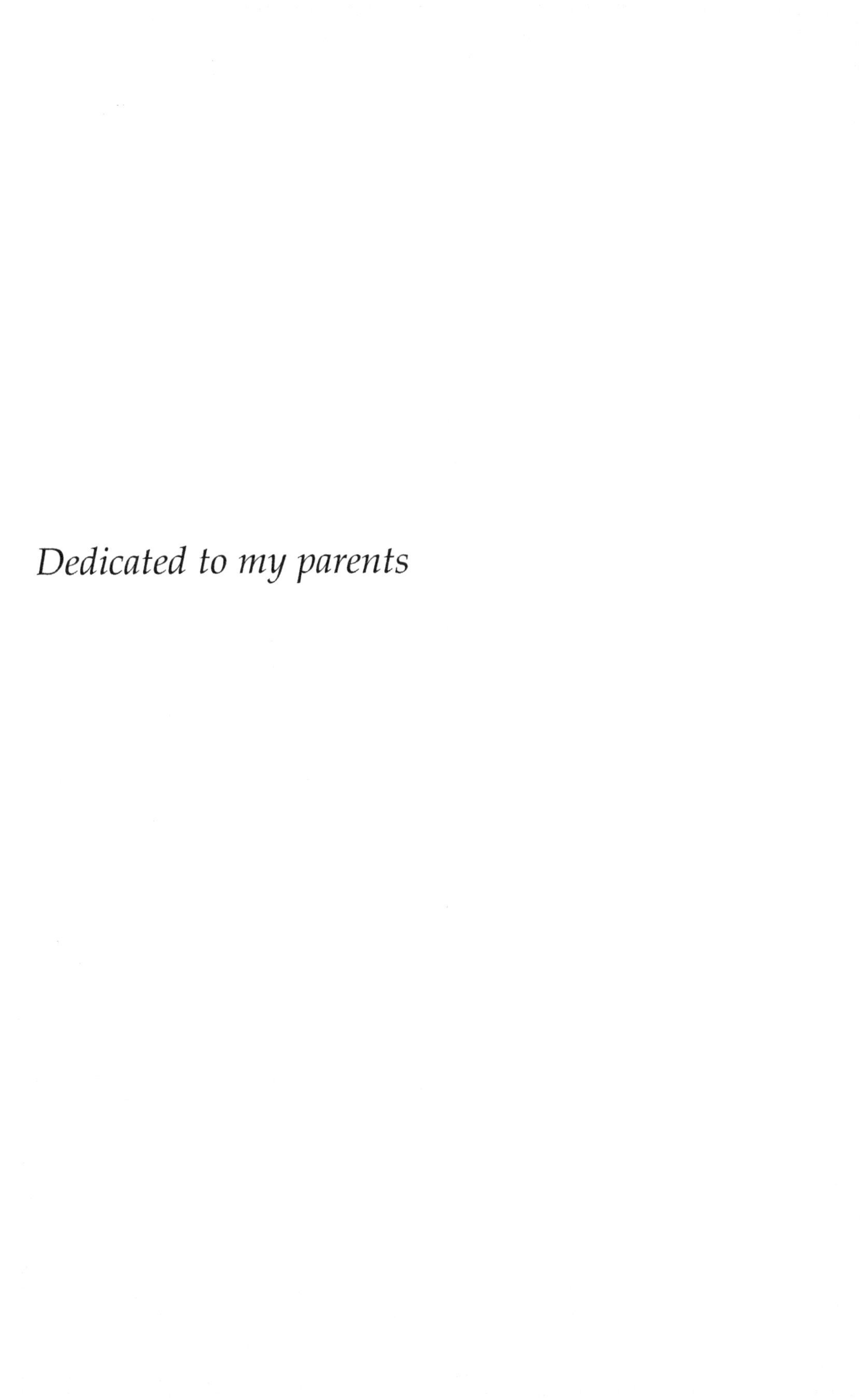

Dedicated to my parents

Preface

There's a deep and beautiful connection between finance and physics. But so far, this connection was used almost exclusively by economists to get a better grip on the financial market.

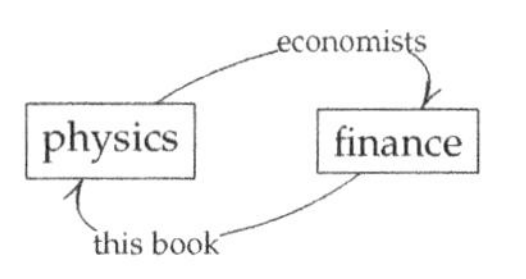

The main idea of this book, however, is to use the connection between finance and physics to translate in the opposite direction. Specifically, the goal of this book is to use a toy version of the financial market to explain fundamental models of nature like general relativity or electrodynamics.[1]

[1] Already at this point, be assured that no previous knowledge of finance theory is necessary to understand this book.

Maybe this sounds like a stupid idea. After all, finance is an extremely difficult subject. Wouldn't it be easier to learn physics directly? Well, I'm convinced that it makes sense to spend a few minutes to think about finance first. Otherwise I wouldn't have spent the time to write this book. But let me explain.

Modern physics often appears like a conglomerate of several seemingly unrelated models.[2] These models are usually discussed completely independently. There are books on general relativity, books on quantum mechanics, books on electrodynamics and students read them in different

[2] Quantum mechanics, general relativity, special relativity, electrodynamics, ...

academic terms.

However, it turns out that the key ideas at the heart of all these models are exactly the same. Therefore, by studying a small number of key ideas, we can understand them all from a common perspective.[3]

Now the bad news is that these key ideas are somewhat abstract. We can't dumb them down to the level of apples and oranges. This is why they are not explained to beginning students and why they are rarely discussed in popular science books.[4] We can, however, understand them by talking about a toy model of the financial market.

This toy model is not trivial and we need some time to think it through. But afterwards, you will not only be able to understand the key ideas at the heart of modern physics but also see the connections between the various fundamental models.

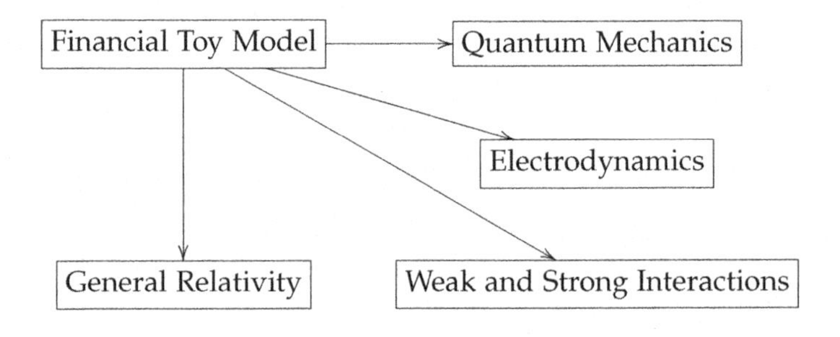

That's why I'm confident that learning a thing or two about the financial market is one of the best investments of your time that you'll ever make.

With all that said, let me make a few things clear.

[3] This is known as the fiber bundle perspective. Fiber bundles are a mathematical concept which provides a unified framework for all modern theories of physics.

[4] Some popular science books try to explain some of these key ideas using analogies. However, these analogies don't represent the main features of the ideas faithfully. This leads to lots of problems when readers try to read more technical texts afterwards. On the other hand, analogies are undoubtedly extremely powerful tools. It can even be argued that all fundamental advances in physics were due to some analogy. (This is discussed extensively in the fantastic book *Surfaces and Essences* by Douglas Hofstadter and Emmanuel Sander.) And, in some sense, this book makes heavy use of analogies, too. But the main difference between the analogies commonly used in popular science books and the finance analogies discussed in this book is that they are, I believe, as close to the real thing as it gets. In particular, this means that you won't run into confusing contradictions when you read more advanced books afterwards.

▷ This is not a book on finance theory. We will not try to build a realistic model of the financial market. Instead, we will only use those parts of the financial market which help us to understand fundamental physics.

▷ After reading this book you won't know everything. There are lots of details missing and many important concepts are not even mentioned. However, this book can do two things for you. First of all, it can help you to see the big picture which often gets lost in more detailed treatments. In other words, you will be able to see the whole forest and not just individual trees. And secondly, it lays the foundation for further studies. By the end of this book you will be equipped with a completely new set of tools which allows you to explore modern physics from a beautiful common perspective.

▷ This is not a popular science book, but it's also not a textbook. It's something in between. While the following chapters contain mathematical equations, the number of mathematical arguments is kept to a minimum and no attempt at mathematical rigor is made. I hope that this way laypersons and students alike can benefit from the following explanations. In particular:

  - If you're a somewhat advanced student of physics, this book allows you to understand many things you probably already know from a completely new perspective.
  - But even if you have no formal physics education, this book will allow you to understand how nature works at a fundamental level. The only prerequisites are a rough understanding of calculus (e.g. what a derivative is) and complex numbers. Undoubtedly, some details will be harder or even impossible to understand without any foreknowledge. But if you feel stuck, you can simply glance over these details and move on. Every important idea will be explained multiple times from various perspectives, with and without equations.

- ▷ I did my best to keep this book as short as possible. In particular, I removed all unnecessary tangents and sections discussing advanced concepts because I believe they would only distract you from the truly fundamental ideas. In addition, I hope the shortness of this book demonstrates that you don't need to spend months of your life to understand the essentials of modern physics.
- ▷ The shortness of this book also means that after reading it you will probably have more questions than you started with. However, I think this is the best thing a book on physics can accomplish because it means your curiosity has been sparked. In Chapter 6, we will discuss in detail which books you should read if you want to dive deeper.

So if you think that this sounds like a solid plan, without any further ado, let's dive in. Hopefully you'll enjoy reading this book as much as I have enjoyed writing it.

Karlsruhe, January 2019 *Jakob Schwichtenberg*

PS: If you find an error, I would appreciate a short email to errors@jakobschwichtenberg.com.

Acknowledgments

Special thanks to Luc Longtin whose comments, ideas and corrections have made this book so much better. Moreover, I want to thank Florian Colbatzky, Andreas Pargner and Jacob Ayres for carefully proofreading the manuscript and Jerry D. Logan, Nam-Kyu Park, Robert Welters and Don Washburn for reporting several typos. I'm also indebted to Paul Tremper for many insightful discussions that inspired me to write this book.

Contents

This page is intentionally left blank

Part I
The Physics of Finance

"We all wish that we had these three laws that explain 99% of all behavior. In fact, economists have 99 laws that explain maybe 3% of economic behavior."

Andrew Lo

1

Finance Intuitively

The main idea of this book is that we can understand many of the most important insights in modern physics as discoveries about the arena modern physics takes place in. In particular, physicists discovered that there is not just spacetime but also *internal* spaces. Spacetime, together with these internal spaces, yields the arena that we use to make sense of modern physics.

Arguably the most important insight was that this arena is not some kind of static background structure but a dynamical part of nature itself. Describing the interplay between elementary particles and this arena is what modern physics is all about.

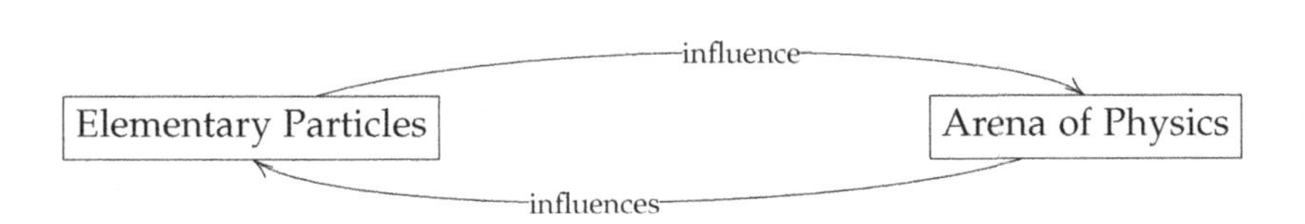

So to understand modern physics we need to answer a few key questions:

▷ What's an internal space?

▷ How can the structure of an internal space be non-trivial?

▷ Where does such a non-trivial structure come from?

▷ How does the non-trivial structure affect objects in the system?

Admittedly all of this sounds horribly abstract.

But, as mentioned already in the preface, we can answer all these questions by thinking about the financial market. This is a useful analogy because, while financial transactions take place at locations in the real world (spacetime), we need abstract internal spaces to keep track of them.

Maybe you don't care about financial markets at all. But let me assure you that it makes sense to think about the following simplified version of the financial market. As soon as we understand how we can describe what is going on here *consistently*, we automatically understand how modern physics works, too.[1]

[1] We will only talk about aspects of the financial market which are directly relevant to modern physics. The following connection between financial markets and fundamental physics was first put forward by Malaney and Weinstein, as summarized in Ref. [Malaney, 1996] and later popularized in Refs. [Ilinski, 1997, Young, 1999, Maldacena, 2016]. The ideas were further developed in [Vazquez and Farinelli, 2009]. After talking about the key concepts in the context of the financial market, we will use them directly to understand how nature works. The main difference between what we do in the financial toy model and in modern physics is that we use different internal spaces.

We will start by talking about symmetries. Symmetries are something which we can grasp intuitively and there is a close connection between the internal spaces and the symmetries of a system.

1.1 Symmetry

First of all, what is a symmetry?

Imagine a friend stands in front of you and holds a perfectly round ball in her hand. Then you close your eyes, your friend performs a transformation of the ball and afterward you open your eyes again. If she rotates the ball while your eyes are closed, it is impossible for you to determine whether or not she did anything at all. Hence, rotations are symmetries of the ball. But if she moves the ball to another location, you will immediately notice it. Therefore, translations are not symmetries of the ball.

Moreover, if she holds a cube, only very special rotations can be done without you noticing it. In general, all transformations which, in principle, change something but lead to an indistinguishable result are symmetries. Formulated differently, a symmetry takes us from one state to a different one, which happens to have the same properties.[2]

[2] In mathematical terms, we usually collect all symmetries of an object and call this set of transformations, together with a rule which allows us to combine them, a **group**. In other words, a group consists of all transformations which leave a specific object unchanged and a rule which allows us to combine two transformations in such a way that we get another symmetry transformation. This is discussed in a bit more detail in Appendix C.

With this in mind, let's talk about an important symmetry of the financial market. Talking about this symmetry will lead us directly to the internal space of our financial toy model.

1.2 Gauge Symmetry

The key observation we need is that prices do not have any absolute meaning. We can shift them arbitrarily, as long as all prices get shifted equally.[3]

[3] Maybe you wonder what happens when different countries use different currencies. Even in that case, prices still have no absolute meaning. We will talk about this in detail below.

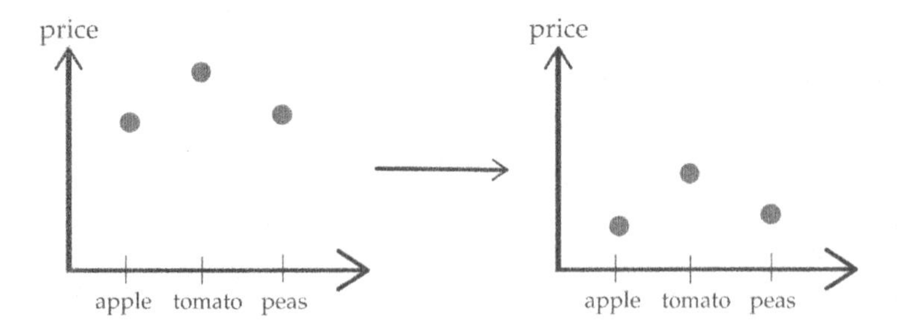

Imagine, for example, that a trader sells three tomatoes at €1 each and then uses this money to buy six apples at €0.50 each.

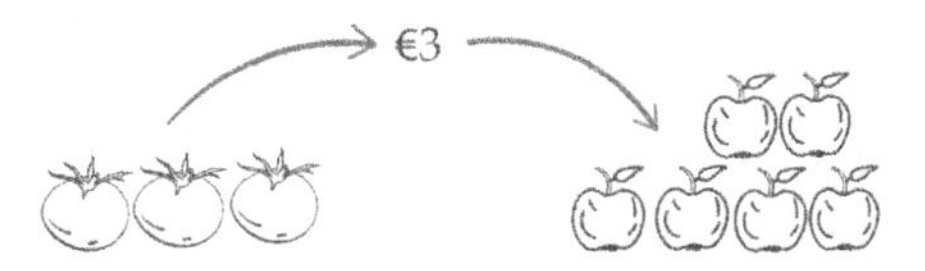

The end result of such a process is completely unchanged if the government decides to print lots of euro notes such that the value of each euro drops by a factor of ten. Afterwards, the trader gets €10 for each tomato but then needs to pay €5 per apple. So again, our trader starts with three tomatoes and ends up with six apples.

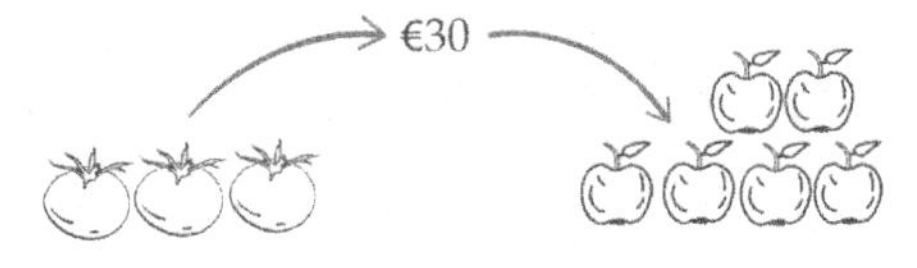

This is the case because fiat money has no fixed absolute value. Thus, we can rescale a given currency without any physical effect.[4]

[4] At least in our toy model changing the value of the currency has no effect, since we imagine that, as a result, all prices and wages are automatically adjusted. Of course, in the real world there could be psychological effects since people get used to certain prices.

In the previous section we learned that whenever we find a transformation which leaves a given object unchanged, we call the transformation a symmetry. In this sense, we say that rescaling a given currency is a symmetry of the financial market.[5]

[5] The freedom to rescale currencies is an example of what we call a **gauge symmetry** in physics. If you want to learn more about gauge symmetries, try *Demystifying Gauge Symmetry* by J. Schwichtenberg available at `https://arxiv.org/abs/1901.10420`.

At first glance, this may seem like an unimportant detail. But we will see below that this symmetry is an extremely useful tool when we want to describe the financial market mathematically.[6]

[6] To spoil the surprise: whenever we want to write down laws describing a system, we know that they must respect its symmetries. In particular, when we want to describe the financial market consistently, we need to take the rescaling symmetry into account. The rescaling of a local currency is not allowed to have any effect on the dynamics of the system. This is a powerful constraint which will allow us to derive the correct equations.

Now, what does this symmetry have to do with the internal space of the financial market?[7]

[7] Recall that one of our main goals in this chapter is to understand what an internal space is.

1.3 Internal Space

For simplicity, let's focus on one specific commodity, say, copper. We can imagine that all possible prices of copper live in an internal space that we call the **money space**. Each point in this space corresponds to one specific price.

But, as we've discussed in the last section, it doesn't matter where exactly we are in money space. We can move the price of copper around freely. The price could be any number between €0.0...01 and €∞.[8] To take all these possibilities into account *simultaneously*, we imagine that there

[8] Here €0.0...01 denotes an arbitrarily small but non-zero positive number.

is an abstract line which represents our money space. Each point on this line corresponds to a specific price for the commodity in the (local) currency. The symmetries of our system are shifts from one point to another.

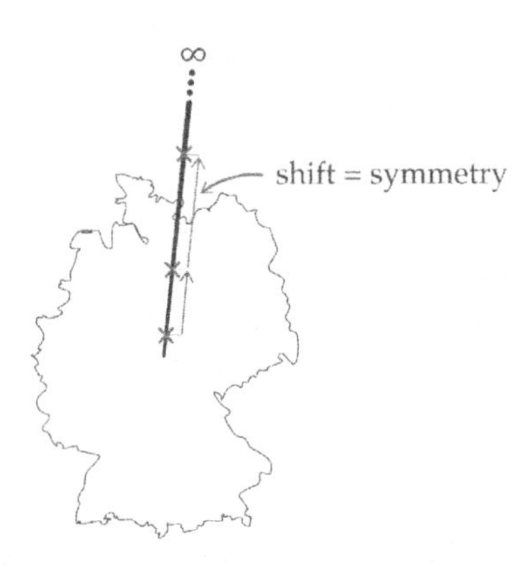

Since different countries possibly use different currencies and have different prices, it's instructive to imagine that there is such a line above each country.[9]

[9] Take note that each line here is, in principle, infinitely long. But, of course, we can't draw that.

Together, all these lines with all possible locations (here all countries) yield the arena which we use to describe what is going on in our toy model.[10]

[10] Don't worry if this construction still seems horribly abstract. Below we will discuss concretely how this picture of the financial market helps us to describe it.

Moreover, take note that this construction is completely analogous to what we do in modern physics. Later we will discuss how and why in physics we also attach internal spaces to each point of spacetime.

Internal spaces are not only important to help us keep track of what is happening but because they can also directly influence the dynamics. To understand this, we need to make our toy model only slightly more realistic.

1.4 Connections

We now need to talk about what happens when countries use different currencies. When we allow that countries use different currencies, it still has to be true that prices alone do not have any meaning. A large number for the price of copper in one country (e.g. 10 Indian rupees per kilogram) and a tiny number in another country (e.g. 0.5 euros per kilogram) does not necessarily mean that copper is cheaper in the latter country.[11]

[11] At the time of writing, January 2019, 10 Indian rupees are worth approximately 0.12 euros.

Whether the price in a given country is cheap or expensive depends on the **exchange rates** of the local currency. If I only tell you that copper trades in Japan for 100 yen per kilogram, it's impossible for you to decide if it makes sense to ship your copper to Japan and sell it there. But if I also tell you that 100 yen are worth 0.8 euros, you know immediately that this is a good price compared to, say, 0.5 euros per kilogram in Germany.[12]

[12] As of January 2019, 1 yen is worth 0.008 euros.

For our geometrical picture, this means that when countries use different currencies we not only need a line above each country but also something which glues them together. Our exchange rates are the glue which allow us to calculate how much a given currency is worth in terms of another currency.

1.5 Gauge Invariant Quantities

All this may seem rather boring or even trivial. But be assured that thoughts like this are really at the heart of modern physics. To understand how this comes about, we need to talk about dynamics.

To describe what is going on in a given system, we need an **equation of motion**. An important idea is that our equation cannot depend on anything which changes if a currency gets rescaled. In particular, this means that our equation of motion cannot depend on the prices of copper directly. These prices depend on the local currency and therefore change as soon as a country rescales its currency.

This is already a powerful insight.[13] Naively, we may have thought that any equation describing our system certainly involves the prices of copper. But we just learned that any such equation wouldn't be useful because prices do not have any direct meaning.

[13] In fact, we will use it below to derive one of the most important equations.

But what does then have an effect on the dynamics?

Well, opportunities to make money certainly have an effect. If such an opportunity shows up somewhere, traders will use it to earn money.

One way to make money is to buy copper in one country and sell it in a different country at a higher price. Spotting these kinds of opportunities is not so simple when countries use different currencies.

In order to be sure that an opportunity is lucrative, we need to be able to calculate whether the value of the currency we will receive when we sell the copper is greater than that of the currency we used to buy the copper. Only when what we finally receive is more valuable than what we started with, do we really earn any money.

For example, if we start by buying copper for 10 euros and end by selling it for 10000 yen, this does not tell us anything unless we calculate how many euros we get for our yen.[14]

[14] As of January 2019, 10000 yen is worth 80.2 euros. So this would be a good deal.

A crucial point is that such an opportunity to earn money does not depend on local conventions. For example, if Japan decides to drop a zero from their currency, you would only get 1000 yen for your copper. But you could still trade these for 80.2 euros, because the euro-yen exchange rate is adjusted as soon as Japan decides to modify its currency.

In physics, we call a quantity which does not depend on

local conventions **gauge invariant**. In our example, the amount of money we can earn through such a copper trade is gauge invariant.

This is an important observation because if we want to write down an equation describing the system, the gain factor which quantifies the opportunity would be a great candidate to appear in it.[15]

[15] We will define the gain factor later in mathematical terms.

What other quantities should appear in our equation of motion? Maybe the exchange rates?

Again, this cannot be the case because exchange rates do depend on local conventions. We can, however, construct a quantity closely related to the exchange rates which does not depend on local conventions and therefore possibly plays a role in our equation of motion.

All opportunities to make money are crucial for the dynamics within our system. So far, we have only talked about how we can earn money by trading copper. It turns out that sometimes it's possible to earn money purely by trading one currency for another. This is known as **currency arbitrage.**[16]

[16] In general, arbitrage describes an opportunity to make a risk-less profit.

For example, let's imagine the exchange rates are as follows:

$$\begin{aligned} DM/P &= 1 \\ P/F &= 2 \\ F/L &= 10 \\ DM/L &= 10, \end{aligned} \tag{1.1}$$

where DM denotes Deutsche Mark, F Francs, P Pounds and L Lira.

A trader is able to earn money by exchanging currencies.

If he starts with $1DM$, he can trade it for $1P$, then use the pound to trade it for $2F$, then trade these for $20L$ and finally trade these back for $2DM$.

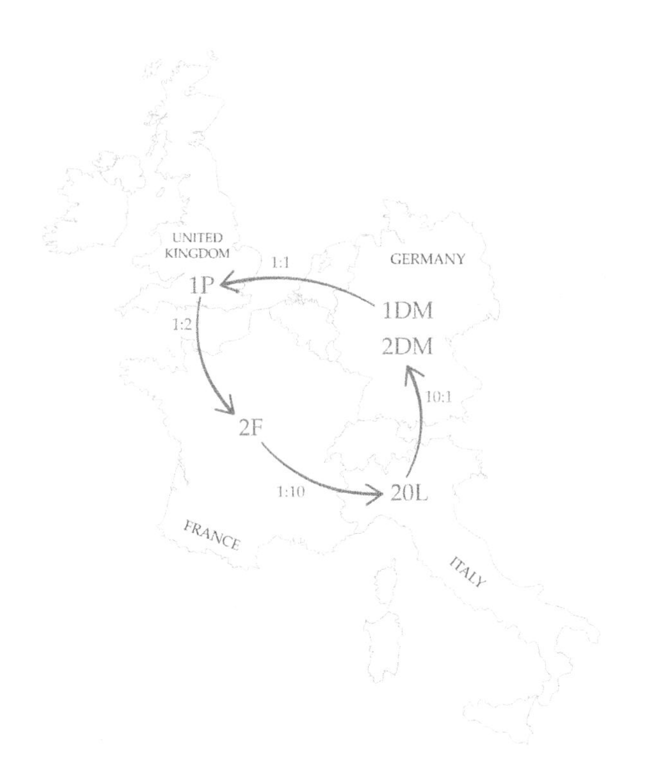

Take note that again it's crucial that we end up with the same currency that we started with because we can't otherwise be sure that we really earned money. Moreover, take note that such an arbitrage possibility is a real feature of the system and does not depend on local conventions.

We can check this explicitly. First of all, our exchange rates do change when, for example, Italy decides to drop a zero from their currency $L \to \tilde{L} = 10L$.[17] As a result of this

[17] We are assuming the financial market works perfectly. Every time a country decides to change its local currency, the exchange rates get adjusted immediately and perfectly. But then, it's natural to wonder: if everything works so perfectly, where do arbitrage opportunities come from? This is something we will talk about below. At this point, our exchange rates are completely passive and their only job is to keep track of local conventions. Later we will introduce new dynamical actors (banks) which adjust the exchange rates *actively*.

change in Italy, our exchange rates change as follows

$$\begin{aligned} DM/P &= 1 \\ P/F &= 2 \\ F/L &= 10 \quad \rightarrow \quad F/\tilde{L} = F/(10L) = 1 \\ DM/L &= 10 \quad \rightarrow \quad DM/\tilde{L} = DM/(10L) = 1\,. \end{aligned} \tag{1.2}$$

While the exchange rates are different, the amount of money our trader earns is unchanged by such a rescaling!

If he starts again with $1DM$, he can still trade it for $1P$, then for $2F$, then trade these for $2\tilde{L}$ and finally trade these for $2DM$. The final result is the same as before.

What we've therefore discovered is a second method to earn money. The gain factor which quantifies how much money we can earn does not depend on local conventions and therefore quite likely plays a dominant role in our equation of motion.[18]

[18] We will define the corresponding gain factor in mathematical terms below.

1.6 Curvature

Before we move on, let's return for a moment to our geometrical picture. Above, I've mentioned that the exchange rates are the glue which ties together the lines representing the local money spaces in the various countries. Our total internal space consists of all these lines glued together. An extremely useful interpretation is that arbitrage opportunities correspond geometrically to a **non-zero curvature** of this internal space.

Intuitively, whenever we move in a loop and end up with a different state than the one we started in, we say the space we move in is curved.[19]

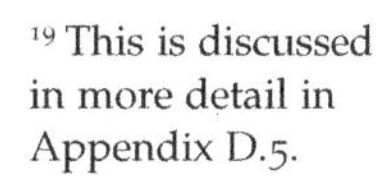
[19] This is discussed in more detail in Appendix D.5.

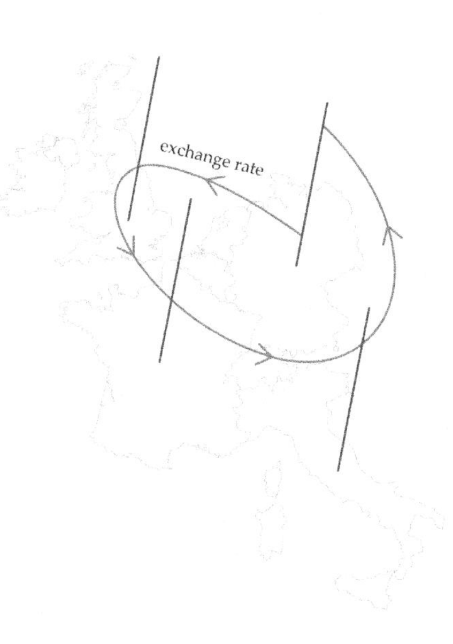

For the moment, simply imagine that if there is an arbitrage possibility, our local internal spaces (lines) are glued together non-trivially. This is a useful picture because whenever a trader performs a financial transaction, he not only moves through spacetime but also through our internal space. His behavior is directly influenced by the curvature of the space, analogous to how a marble rolls differently on a curved surface.

Before we move on and discuss things like this in detail, there is one more feature of the toy model that we need to discuss.

1.7 Dynamical Curvature

In a real financial market there are, of course, not only traders but also other actors. Most importantly, there are banks which determine the exchange rates and allow traders to exchange one currency for another. But these banks are not purely sitting in the background passively. Instead, they are dynamically shaping what is going on in the system. In particular, they adjust the exchange rates all the time. They do this, for example, to get rid of arbitrage opportunities because otherwise they would quickly run out of money.

This means that there is a continuous interplay between the two kinds of actors. The behavior of traders is directly influenced by the pattern of arbitrage opportunities within the system. In turn, the banks modify the exchange rates all the time depending on the behavior of the traders.[20]

[20] We will make this more concrete below.

This idea allows us to get a first glimpse at what all this has to do with physics.

First of all, in physics, we follow exactly the same steps to find quantities which do not depend on local conventions. With these quantities at hand, writing down the equations governing the dynamics of the system is only a small additional step. In addition, as we will discuss later, the gain factors we discussed in the context of the financial toy model are *mathematically* completely analogous to the quantities used to describe fundamental physics![21] Moreover, the interplay between traders and exchange rates is completely analogous to the interplay between elementary particles and force fields.[22]

[21] In mathematical terms, the gain factors describing currency arbitrage opportunities represent the curvature of our internal space. In physics, we call the curvature (at least if it's a dynamical actor) a gauge field.

[22] Examples of force fields are the gravitational field and the electromagnetic field.

Before we start talking about physics, it is extremely instructive to rephrase everything that we have just learned in mathematical terms. This allows us to understand many of the most important concepts of modern physics and to write down one of the most important equations.

2

Finance Mathematically

First of all, a short warning: It's easily possible that you will get bored or confused by reading the following sections. But please don't let that discourage you from reading the rest of the book. When you feel lost, simply skim through the remaining sections and then move on to the next chapter. You will be able to understand the rest of the book even if you don't internalize all the mathematical details in this chapter. Moreover, you can always come back later and catch up on the details.

In addition, take note that this section is inspired by the excellent essay titled "The symmetry and simplicity of the laws of physics and the Higgs boson" by Juan Maldacena.[1] *Therefore, if you don't understand one of the explanations below, you can consult this essay to read a second perspective.*

[1] The essay is available at `https://arxiv.org/abs/1410.6753`.

With this in mind, let's dive in.

Mathematically, we imagine that our **countries** live on a two-dimensional lattice. Each point on the lattice is labelled by 2-numbers: $\vec{n} = (n_1, n_2)$.

In other words, each country can be identified by a vector $\vec{n}$ which points to its location.

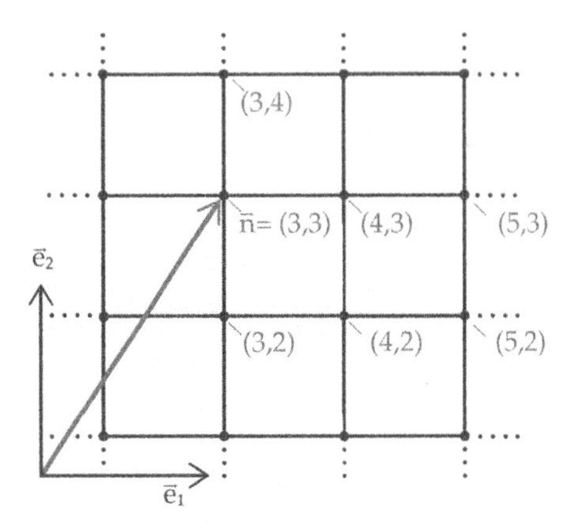

We can move from one country to a neighboring country by using a **basis vector** $\vec{e}_i$, where i denotes the direction we are moving in. For example, $\vec{e}_1 = (1, 0)$.

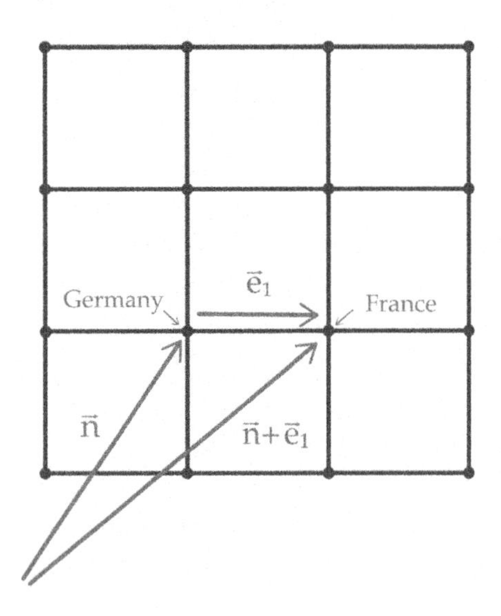

We denote the **exchange rates** between the country labelled by the vector $\vec{n}$ and its neighbor in the i-direction by $R_i(\vec{n})$. For example, if the country at the location labelled by $\vec{n}$ uses Deutsche Marks and its neighbor in the 1-direction uses Francs, $R_1(\vec{n})$ tells us how many Francs we get for each Deutsche Mark.

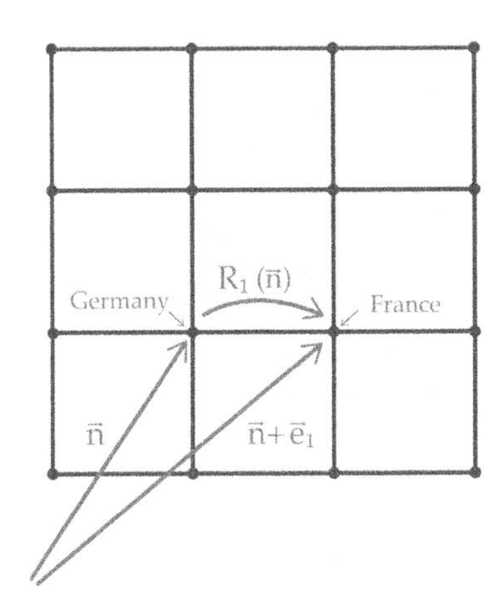

Analogously, $R_2(\vec{n})$ describes the exchange rate between the country at $\vec{n}$ and the country at $\vec{n} + \vec{e}_2$:

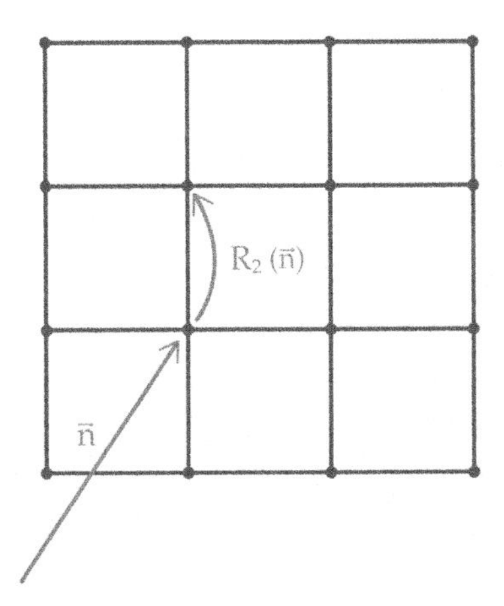

And $R_1(\vec{n} + \vec{e}_2)$ describes the exchange rate between the country at $\vec{n} + \vec{e}_2$ and the country at $\vec{n} + \vec{e}_2 + \vec{e}_1$:

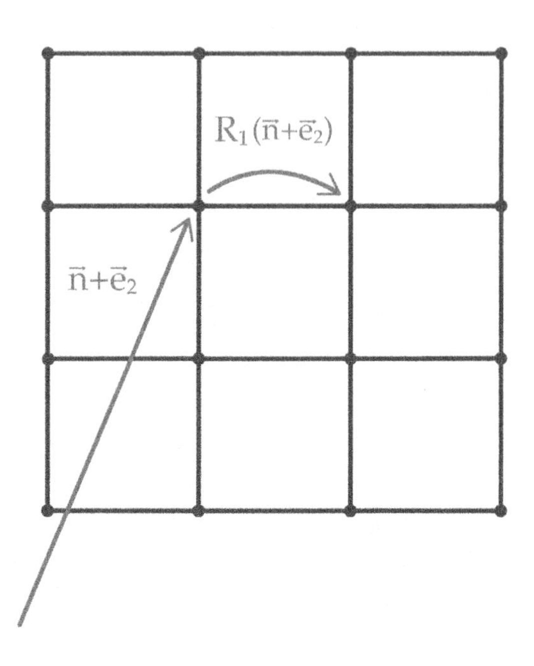

In physics, we usually introduce the corresponding logarithms of the exchange rates[2]

$$R_i(\vec{n}) \equiv e^{A_i(\vec{n})}, \tag{2.1}$$

where $A_i(\vec{n}) \equiv \ln\Big(R_i(\vec{n})\Big)$.[3]

[2] We will see below that this makes our formulas much more simple. In particular, the products in the transformation rules can be replaced by sums of the logarithms. In mathematical terms what we do here is to work with the Lie algebra instead of with the corresponding Lie group. While a Lie group is, in general, a complicated object, the corresponding Lie algebra is always a simple vector space. Lie algebras are discussed in Appendix C.3.

[3] The natural logarithm is the inverse of the exponential function. This means,

$$e^{\ln(x)} = x.$$

The next ingredient that we need is a notation for gauge transformations. In our toy model a gauge transformation is a rescaling of a local currency and directly impacts, for example, the exchange rates. We use the notation $f(\vec{n})$ to denote a rescaling of the currency in the country at $\vec{n}$ by a factor f.

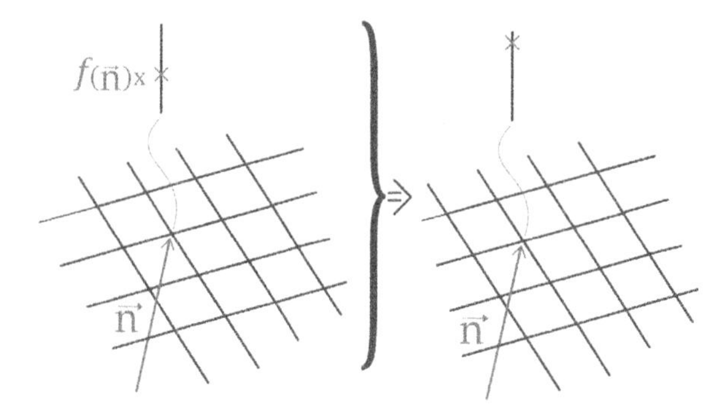

In addition, we again introduce the corresponding logarithm

$$f(\vec{n}) \equiv e^{\epsilon(\vec{n})} . \tag{2.2}$$

In general, when we perform such a gauge transformation, $f(\vec{n})$, in the country labelled by $\vec{n}$ and also another gauge transformation in the neighboring country in the i-direction, $f(\vec{n} + \vec{e}_i)$, the corresponding exchange rate gets modified by the ratio of these two factors:

$$R_i(\vec{n}) \rightarrow \frac{f(\vec{n} + \vec{e}_i)}{f(\vec{n})} R_i(\vec{n}) . \tag{2.3}$$

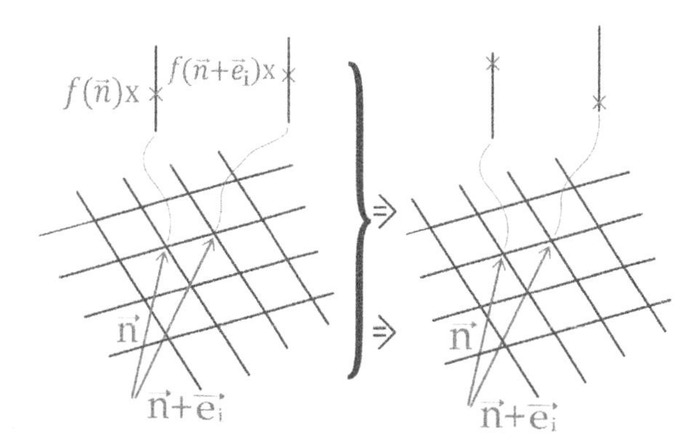

In terms of the logarithms this equation reads

$$\begin{aligned} R_i(\vec{n}) \overset{\text{Eq. 2.1}}{=} e^{A_i(\vec{n})} &\to \frac{f(\vec{n}+\vec{e}_i)}{f(\vec{n})} R_i(\vec{n}) \\ &\quad \text{Eq. 2.2} \\ &= \frac{e^{\epsilon(\vec{n}+\vec{e}_i)}}{e^{\epsilon(\vec{n})}} e^{A_i(\vec{n})} \\ &\quad \frac{e^A}{e^B} = e^{A-B} \\ &= e^{A_i(\vec{n})+\epsilon(\vec{n}+\vec{e}_i)-\epsilon(\vec{n})} \end{aligned} \tag{2.4}$$

and we can conclude that

$$A_i(\vec{n}) \to A_i(\vec{n}) + \epsilon(\vec{n}+\vec{e}_i) - \epsilon(\vec{n}) \,. \tag{2.5}$$

We learned above that an important aspect of the system is whether arbitrage opportunities exist. An arbitrage possibility exists when we can trade currencies in such a way that we end up with more money than we started with. But we can only make such a statement when the initial currency and the final currency are the same. Only then can we be certain whether or not the final amount of money is larger than the amount we started with. Therefore, we need to trade money in a loop.

Defining the **gain factor** as the ratio of the final amount of money to the initial amount of money (in the same local currency), the total gain we can make by following a specific loop can be quantified by[4]

[4] We will discuss why this formula is correct below.

$$G(\vec{n}) = R_i(\vec{n}) R_j(\vec{n}+\vec{e}_i) \frac{1}{R_i(\vec{n}+\vec{e}_j)} \frac{1}{R_j(\vec{n})} \,. \tag{2.6}$$

When this gain factor is larger than one, we can earn money by trading money following the loop. If it is smaller than one, we lose money.

To understand Eq. 2.6 for the gain factor in the case of a loop process, imagine that we start with $1DM$. We trade it for Pounds and $R_1(\vec{n}) = 1\frac{\text{P}}{\text{DM}}$ tells us that we get in total $1P$. Afterwards, we trade our Pounds for Francs and $R_2(\vec{n}+\vec{e}_1) = 2\frac{\text{F}}{\text{P}}$ tells us that we get in total $2F$. Afterwards, we trade our Francs for Lira. $R_1(\vec{n}+\vec{e}_2) = \frac{1}{10}\frac{\text{F}}{\text{L}}$ tells us that we get 10 Lira for each Franc. Hence, we have to calculate $2F/R_1(\vec{n}+\vec{e}_2) = 20L$. Finally, we use that $R_2(\vec{n}) = 10\frac{\text{L}}{\text{DM}}$ tells us that we need to pay $10L$ for each Deutsche Mark and therefore calculate $20L/R_2(\vec{n}) = 2DM$.[5] The gain factor in this case is:

$$\frac{\text{final amount}}{\text{initial amount}} = \frac{2\text{DM}}{1\text{DM}} = 2, \tag{2.7}$$

which is what is obtained by substituting values for the exchange rates in Eq. 2.6.

[5] The crucial point is that our exchange rates $R_i(\vec{n})$ always tell us how much of the currency in the neighboring country in the i-direction we get for each unit of the local currency in the country at $\vec{n}$. Hence, we sometimes have to divide by the corresponding exchange rate to calculate the resulting amount of a new currency.

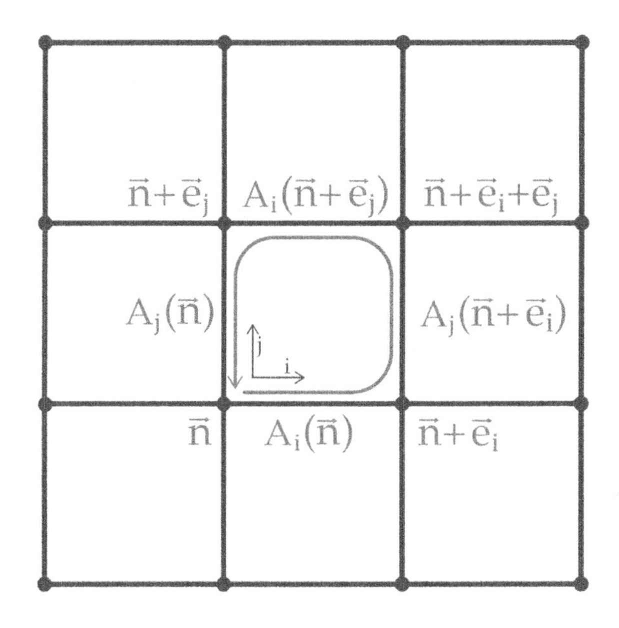

Once more we introduce the corresponding logarithm

$$G(\vec{n}) \equiv e^{F_{ij}(\vec{n})} \tag{2.8}$$

and again, we can rewrite our equation (Eq. 2.6) in terms of the logarithms[6]

[6] To spoil the surprise: in physics F_{ij} is directly related to components of the **magnetic field**. For example, $F_{12} = -B_3$.

$$
\begin{aligned}
G(\vec{n}) &= R_i(\vec{n})R_j(\vec{n}+\vec{e}_i)\frac{1}{R_i(\vec{n}+\vec{e}_j)}\frac{1}{R_j(\vec{n})} && \text{Eq. 2.8, Eq. 2.1} \\
e^{F_{ij}(\vec{n})} &= e^{A_i(\vec{n})}e^{A_j(\vec{n}+\vec{e}_i)}\frac{1}{e^{A_i(\vec{n}+\vec{e}_j)}}\frac{1}{e^{A_j(\vec{n})}} && \frac{e^A}{e^B}=e^{A-B} \\
&= e^{A_i(\vec{n})+A_j(\vec{n}+\vec{e}_i)-A_i(\vec{n}+\vec{e}_j)-A_j(\vec{n})}\,. && (2.9)
\end{aligned}
$$

We can read off here that

$$
F_{ij}(\vec{n}) = A_j(\vec{n}+\vec{e}_i) - A_j(\vec{n}) - \left[A_i(\vec{n}+\vec{e}_j) - A_i(\vec{n})\right]. \quad (2.10)
$$

A crucial consistency check is that G and F_{ij} are unchanged by gauge transformations. An arbitrage possibility is something real and thus cannot depend on local choices of the money internal system. Quantities like this are usually called gauge invariant. So in words, G and F_{ij} encode what is physical in the structure of exchange rates.[7] Moreover, an important technical observation is that $F_{ij}(\vec{n})$ is antisymmetric: $F_{ij}(\vec{n}) = -F_{ji}(\vec{n})$, which follows directly from the definition (Eq. 2.9).[8]

[7] A single (logarithm of an) exchange rate $A_i(\vec{n})$ is gauge dependent and can therefore be set to zero simply by changing local money coordinate systems.

[8] Antisymmetry means that we get a minus sign when we exchange the indices $i \leftrightarrow j$. Using Eq. 2.9, we can calculate

$$
\begin{aligned}
F_{ji}(\vec{n}) &= \\
&A_i(\vec{n}+\vec{e}_j) - A_i(\vec{n}) \\
&\quad - [A_j(\vec{n}+\vec{e}_i) - A_j(\vec{n})] \\
&= -\Big(A_j(\vec{n}+\vec{e}_i) - A_j(\vec{n}) \\
&\quad - [A_i(\vec{n}+\vec{e}_j) - A_i(\vec{n})]\Big) \\
&= -F_{ij}\,.
\end{aligned}
$$

So far, we have only talked about spatial exchange rates. However, there are also temporal exchange rates, i.e. interest rates. A clever trick to incorporate this is to introduce time as the zeroth-coordinate like we do in special relativity.[9] In other words, in addition to specific locations (countries) our lattice contains copies of these locations at different points in time.

[9] We will talk about special relativity in Section 5.1.

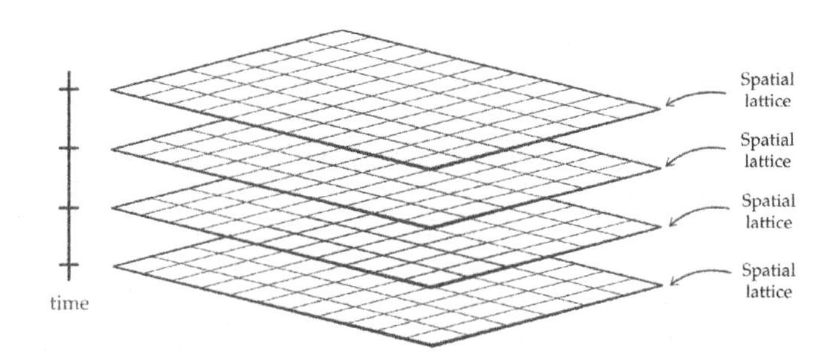

This means a point on this spacetime lattice is specified by $2+1$ coordinates $\vec{n} = (n_0, n_1, n_2)$ and the zeroth component indicates the point in time.

Then, Eq. 2.9 reads[10]

$$F_{\mu\nu}(\vec{n}) = A_\nu(\vec{n}+\vec{e}_\mu) - A_\nu(\vec{n}) - [A_\mu(\vec{n}+\vec{e}_\nu) - A_\mu(\vec{n})], \quad (2.11)$$

where previously $ij \in \{1,2\}$ and now $\mu, \nu \in \{0,1,2\}$.

[10] Once more, let me spoil the surprise: the non-vanishing components of $F_{\mu\nu}(\vec{n})$ with either $\mu = 0$ or $\nu = 0$ are directly related to what we call the electric field in physics. For example, $F_{01} = E_1$. An important difference is that in physics, we're typically dealing with three spatial dimensions and hence the indices μ, ν run from 0 to 3 and the corresponding spatial indices, i, j run from 1 to 3.

An important step to make our model more realistic is to allow not just one possible location per country, but in principle, infinitely many. Graphically, we can imagine that we add more and more points to our lattice. These points represent possible locations of our traders.

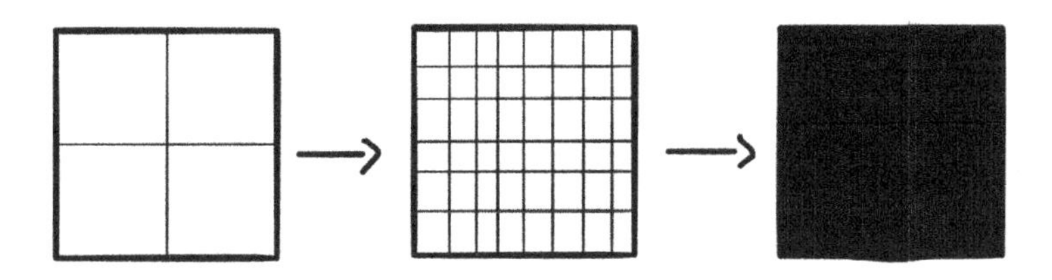

Each possible location corresponds to a vector $\vec{x} = a\vec{n}$, where $\vec{n}$ is a vector with integer components, $\vec{n} = (n_i, n_j)^T$. These integers get larger and larger in value, at any specific location, as a gets smaller and smaller. This is because we fit in more and more points (locations) within a given space interval. However, the length of a vector $\vec{x}$ remains fixed and finite in size. Usually, a is known as the **lattice spacing**. The limit, in which there is a continuum of possible locations, is known as the **continuum limit**.

With this notation at hand, we can specify what the continuum limit means mathematically. In the continuum limit, the distance between neighboring lattice points goes to zero $a \to 0$ while $\vec{x}$ has to remain a fixed finite vector.[11] Moreover, since our bookkeepers $A_i(\vec{n})$ connect lattice points on our discrete lattice, we need to replace them with bookkeepers which connect points in our continuous space $A_i(\vec{n}) \to a\tilde{A}_i(\vec{x})$. As we will see in a moment, the factor a makes sure that the values of our bookkeepers stay fixed, as we take the limit $a \to 0$.

[11] Take note that mathematically this implies that $\vec{n}$ goes to infinity. But this is nothing you should worry about because the only thing we care about after we've taken the limit is a finite $\vec{x}$.

So, specifically, the transformation in Eq. 2.5

$$A_i(\vec{n}) \to A'_i(\vec{n}) \equiv A_i(\vec{n}) + \epsilon(\vec{n}+\vec{e}_i) - \epsilon(\vec{n}) \tag{2.12}$$

now reads

$$\begin{aligned} a\tilde{A}_i(\vec{x}) \to a\tilde{A}'_i(\vec{x}) &\equiv a\tilde{A}_i(\vec{x}) + \epsilon(\vec{x}+a\vec{e}_i) - \epsilon(\vec{x}) \qquad \text{ ⤵ } \frac{a}{a}=1 \\ &= a\tilde{A}_i(\vec{x}) + a\frac{\epsilon(\vec{x}+a\vec{e}_i) - \epsilon(\vec{x})}{a} \qquad \text{ ⤵ } \\ &= a\left(\tilde{A}_i(\vec{x}) + \frac{\epsilon(\vec{x}+a\vec{e}_i) - \epsilon(\vec{x})}{a}\right). \end{aligned} \tag{2.13}$$

In the limit $a \to 0$, the second term on the right-hand side yields a difference quotient[12]

[12] This is how the derivative of a function is defined in general.

$$\lim_{a\to 0} \frac{\epsilon(\vec{x}+a\vec{e}_i) - \epsilon(\vec{x})}{a} \equiv \frac{\partial \epsilon}{\partial x_i}. \tag{2.14}$$

And by looking at Eq. 2.13 we can therefore conclude that[13]

[13] We can now see why we introduced the additional factor a in the definition of $\tilde{A}_i$. This factor makes sure that a drops out completely from our equation and we get a quantity which remains finite in the limit $a \to 0$.

$$\tilde{A}_i(\vec{x}) \to \tilde{A}'_i(\vec{x}) \equiv \tilde{A}_i(\vec{x}) + \frac{\partial \epsilon}{\partial x_i}. \tag{2.15}$$

Analogously, the gain factor in Eq. 2.9,

$$F_{ij}(\vec{n}) = A_j(\vec{n}+\vec{e}_i) - A_j(\vec{n}) - [A_i(\vec{n}+\vec{e}_j) - A_i(\vec{n})], \tag{2.16}$$

becomes

$$F_{ij}(\vec{x}) = a\tilde{A}_j(\vec{x}+a\vec{e}_i) - a\tilde{A}_j(\vec{x}) - [a\tilde{A}_i(\vec{x}+a\vec{e}_j) - a\tilde{A}_i(\vec{x})]$$

$$\curvearrowleft \quad \frac{a}{a} = 1$$

$$= a^2\frac{\tilde{A}_j(\vec{x}+a\vec{e}_i) - \tilde{A}_j(\vec{x})}{a} - a^2\frac{\tilde{A}_i(\vec{x}+a\vec{e}_j) - \tilde{A}_i(\vec{x})}{a} \tag{2.17}$$

and we can conclude that we have to replace $F_{ij} \to a^2\tilde{F}_{ij}$ to get a quantity with a fixed value in the limit $a \to 0$

$$a^2\tilde{F}_{ij}(\vec{x}) = a^2\frac{\tilde{A}_j(\vec{x}+a\vec{e}_i) - \tilde{A}_j(\vec{x})}{a} - a^2\frac{\tilde{A}_i(\vec{x}+a\vec{e}_j) - \tilde{A}_i(\vec{x})}{a}$$

$$\curvearrowleft \quad \not{a}^2$$

$$\tilde{F}_{ij}(\vec{x}) = \frac{\tilde{A}_j(\vec{x}+a\vec{e}_i) - \tilde{A}_j(\vec{x})}{a} - \frac{\tilde{A}_i(\vec{x}+a\vec{e}_j) - \tilde{A}_i(\vec{x})}{a} \tag{2.18}$$

On the right-hand side, we again have difference quotients in the continuum limit

$$\lim_{a\to 0}\frac{\tilde{A}_j(\vec{x}+a\vec{e}_i) - \tilde{A}_j(\vec{x})}{a} = \frac{\partial}{\partial x_i}\tilde{A}_j$$
$$\lim_{a\to 0}\frac{\tilde{A}_i(\vec{x}+a\vec{e}_j) - \tilde{A}_i(\vec{x})}{a} = \frac{\partial}{\partial x_j}\tilde{A}_i\,. \tag{2.19}$$

We therefore find that

$$\tilde{F}_{ij}(\vec{x}) = \frac{\partial}{\partial x_i}\tilde{A}_j - \frac{\partial}{\partial x_j}\tilde{A}_i\,. \tag{2.20}$$

Moving forward, we will drop the tilde above our symbols to unclutter the notation. Moreover, if we include temporal exchange rates, we have to replace our indices i, j with μ, ν which each run from 0 to 2.

To summarize, in the continuum limit Eq. 2.5 becomes

$$A_\mu(t,\vec{x}) \to A_\mu(t,\vec{x}) + \frac{\partial \epsilon}{\partial x_\mu} \tag{2.21}$$

and Eq. 2.11 becomes

$$F_{\mu\nu}(t,\vec{x}) \equiv \frac{\partial A_\nu}{\partial x_\mu} - \frac{\partial A_\mu}{\partial x_\nu}. \tag{2.22}$$

In the previous chapter we noticed that we can not only earn money by trading currencies, but also by trading goods like copper. Depending on the local prices it can be lucrative to buy copper in one country, bring it to another country, sell it there, and then go back to the original country to compare the final amount of money with the amount of money we started with.

From the definition of the gain factor as the ratio of the final amount of money to the initial amount of money (in the same local currency), the gain factor for such a process is given by

$$g_i(\vec{n}) = \frac{p(\vec{n}+\vec{e}_i)}{p(\vec{n})R_i(\vec{n})}, \tag{2.23}$$

where $p(\vec{n})$ denotes the price of copper in the country located at $\vec{n}$. To understand the definition in Eq. 2.23, imagine that we start with $10DM$ and the price for one kilogram of copper in Germany is $p(\vec{n}) = 10DM$. This means that we can buy exactly 1 kilogram of copper. Then we can go to the neighboring country and sell our copper for, say, $30F$ since $p(\vec{n}+\vec{e}_1) = 30F$. Afterward, we can go back to Germany and exchange our $30F$ for $15DM$ since, say, $R_i(\vec{n}) = 2$. Therefore, we will have made $5DM$ in total; and so the gain factor in this case is:

$$\frac{\text{final amount}}{\text{initial amount}} = \frac{15DM}{10DM} = 1.5. \tag{2.24}$$

This is exactly what is obtained by substituting values for the prices and exchange rate in Eq. 2.23. Again, a gain factor larger than one means that we earn money and a gain factor smaller than one means that we lose money.[14]

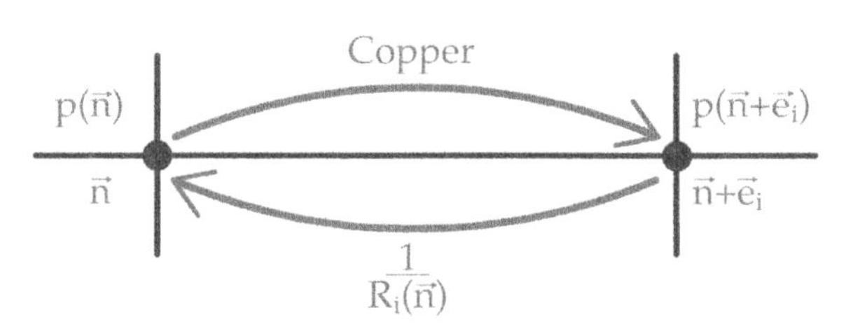

Once more, we introduce the corresponding logarithms

$$g_i(\vec{n}) \equiv e^{H_i(\vec{n})}$$
$$p(\vec{n}) \equiv e^{\varphi(\vec{n})} \tag{2.25}$$

and Eq. (2.23) then reads in terms of the corresponding logarithms

$$g_i(\vec{n}) = \frac{p(\vec{n}+\vec{e}_i)}{p(\vec{n})R_i(\vec{n})}$$

↷ Eq. 2.25, Eq. 2.1

$$e^{H_i(\vec{n})} = \frac{e^{\varphi(\vec{n}+\vec{e}_i)}}{e^{\varphi(\vec{n})}e^{A_i(\vec{n})}}$$

↷

$$= e^{\varphi(\vec{n}+\vec{e}_i)-\varphi(\vec{n})-A_i(\vec{n})}, \tag{2.26}$$

which implies

$$H_i(\vec{n}) = \varphi(\vec{n}+\vec{e}_i) - \varphi(\vec{n}) - A_i(\vec{n}). \tag{2.27}$$

The gain factor $g_i(\vec{n})$ and its logarithm $H_i(\vec{n})$ quantify directly how lucrative copper trade is for a trader based at location $\vec{n}$ and doing business with the country at $\vec{n}+\vec{e}_i$. We can imagine that the more lucrative a trade is, the more will traders make use of it to earn money.

This motivates us to define the **copper current**

$$J_i(\vec{n}) \equiv qH_i(\vec{n}) = q\Big(\varphi(\vec{n}+\vec{e}_i) - \varphi(\vec{n}) - A_i(\vec{n})\Big), \tag{2.28}$$

[14] If you are unsure which quantity goes in the numerator and which in the denominator, ask yourself: would it increase our profit if the given quantity is larger? We write in the numerator if the answer is yes. If the answer is no, we write in the denominator. For example, a higher price of copper in France certainly increases our profit. Hence $p(\vec{n}+\vec{e}_i)$ is written in the numerator. Similarly, a higher price of copper in Germany would lower our profit and therefore, we write in the denominator.

where q is a proportionality constant that quantifies how strongly the traders in a given system react to an opportunity to earn money by trading copper.

Completely analogously to what we did above, we can generalize this formula for situations that involve time by replacing $i \in \{1,2\}$ with $\mu \in \{0,1,2\}$):

$$J_\mu(\vec{n}) = q\Big(\varphi(\vec{n}+\vec{e}_\mu) - \varphi(\vec{n}) - A_\mu(\vec{n})\Big)\,. \tag{2.29}$$

Moreover, in the continuum limit Eq. 2.29 becomes

$$J_\mu(t,\vec{x}) = q\Big(\frac{\partial}{\partial x_\mu}\varphi(t,\vec{x}) - A_\mu(t,\vec{x})\Big)\,. \tag{2.30}$$

Let's talk about the three quantities $J_\mu(\vec{n})$ (with $\mu \in \{0,1,2\}$) in more detail.

The gain factor $H_i(\vec{n})$ tells us how lucrative a trade is in which copper is transported from the country at $\vec{n}$ to the neighboring country at $\vec{n}+\vec{e}_i$. Hence, $J_i(\vec{n})$ is a measure of the amount of copper that flows between the two countries. The trade related to $H_0(\vec{n})$ does not involve the exchange of copper between neighboring countries. Instead, $H_0(\vec{n})$ tells us how much we can earn by buying copper and selling it at a later point in time in the *same* country. Therefore, $J_0(\vec{n})$ is a measure of the amount of copper in the country at $\vec{n}$.

This is an important idea since, so far, we had nothing in our description that contained any information about the amount of copper at a certain location or about how copper flows through the system. Copper is represented in our description in somewhat abstract terms by its price. However, the amount of copper in a given country is not proportional to the local price of copper. The local price depends on the local money coordinate system and therefore cannot directly represent a measurable quantity like the amount of copper. Instead, the amount of copper in a given country must be represented by a gauge independent quantity like $J_0(\vec{n})$. The same is true for the flow of copper $J_i(\vec{n})$.

Maybe a second perspective helps to understand this better. The main idea is that the distribution and flow of copper is somehow encoded in the pattern of prices in the various countries at different times. The prices themselves contain no information about the flow and distribution of copper since they depend on local conventions. Instead, everything we need to know is encoded in how prices change as we move from country to country or to a different point in time.

For example, if the price in a given country goes down rapidly over time, this signals that they have "too much" copper. In other words, a rapidly sinking price is an indicator that lots of copper is located in the corresponding country. The gain factor $H_0(\vec{n})$ describes accurately how the price in the country located at $\vec{n}$ changes over time.[15] Using these ideas, we defined $J_0(\vec{n}) \equiv qH_0(\vec{n})$ which describes how much copper is located in the country at $\vec{n}$. The proportionality constant q is necessary to describe how given amounts of copper (e.g., 3 kilograms or 2 cubic meters) are related to specific changes of the price $H_0(\vec{n})$.

[15] As discussed above, we can't compare the prices directly but need to take into account that price differences can be the result of changes in the currency. For this reason, the definition of $H_0(\vec{n})$ contains an additional term $-A_0(\vec{n})$ which adjusts the price differences appropriately.

Similarly, if the price in one country is much higher than in a neighboring country, we can assume that they desperately need cooper and hence that there is net influx of copper. Therefore, a price difference in neighboring countries is an indicator that copper flows from the low-price country to the high-price country. The price difference between the country at $\vec{n}$ and its neighbor at $\vec{n} + \vec{e}_i$ is described accurately by $H_i(\vec{n})$.[16] Therefore, we use $J_i(\vec{n}) \equiv qH_i(\vec{n})$ to describe the flow of copper from $\vec{n}$ to $\vec{n} + \vec{e}_i$.

[16] Once more, we need to take into account that price differences can be the result of different local conventions. That's why the definition of $H_i(\vec{n})$ contains an additional factor $-A_i(\vec{n})$.

Now that we've introduced a mathematical notation to describe our toy model, we can also write down specific rules which describe how the system behaves. We need two kinds of rules. On the one hand, we need rules which describe how arbitrage opportunities show up and evolve over time. On the other hand, we need rules which describe how our traders behave and in particular, how they react to the presence of arbitrage opportunities. The following diagram summarizes what we will discover in the following sections.

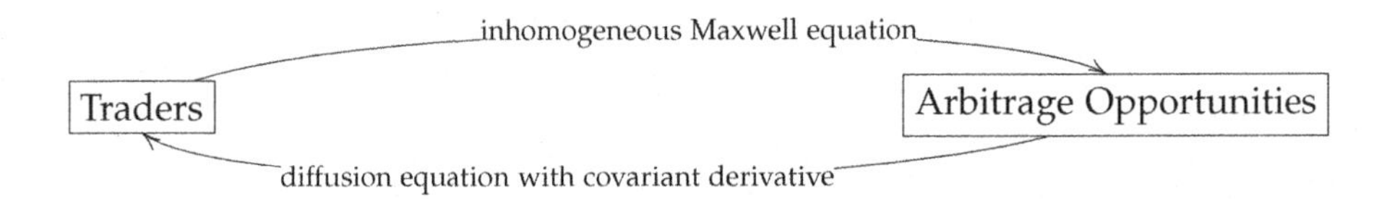

We will start with the rules describing arbitrage opportunities (the inhomogeneous Maxwell equation) and then move on to derive the rules which describe how traders react to them (the diffusion equation with covariant derivative).

2.1 Arbitrage Dynamics

Of course, it's possible to consider various kinds of dynamics within our financial toy model. Different kinds of dynamics require different kinds of laws. However, in the following we discuss a very particular set of rules which correspond to what are known as Maxwell's equations. So, as mentioned in the preface, our goal is not to derive equations which describe financial markets accurately, but to use the financial market to understand concepts like Maxwell's equations.

To derive these, we start with the crucial assumption that copper is conserved, i.e. no copper is destroyed or produced. In particular, this means that whenever the amount of copper decreases in a given country it must have gone somewhere. Conversely, whenever the amount of copper increases in a country it must have come from somewhere.[17]

In mathematical terms this means:[18]

$$\text{change of copper in country } \vec{n} = \text{ total net flow}$$

which we can write in mathematical terms as

$$J_0(\vec{n}+\vec{e}_0) - J_0(\vec{n}) = -\left(\sum_{i=1}^{2} J_i(\vec{n}) - \sum_{i=1}^{2} J_i(\vec{n}-\vec{e}_i)\right)$$

where we sum over all neighboring countries. We can also write this as

$$J_0(\vec{n}+\vec{e}_0) - J_0(\vec{n}) + \left(\sum_{i=1}^{2} J_i(\vec{n}) - \sum_{i=1}^{2} J_i(\vec{n}-\vec{e}_i)\right) = 0\,. \quad (2.31)$$

and we can visualize this equation as follows:

[17] Additionally, take note that no teleportation of copper is allowed.

[18] We only consider elementary copper trade loops. In these elementary loops copper always flows "from left to right". For example, $J_i(\vec{n})$ is proportional to the amount of copper that flows from $\vec{n}$ to $\vec{n}+\vec{e}_i$. A positive $J_i(\vec{n})$ therefore represents an outflow of copper. Analogously, the flow involved in the process which yields $J_i(\vec{n}-\vec{e}_i)$ also goes from left to right. However, a positive $J_i(\vec{n}-\vec{e}_i)$ means a net inflow. Therefore, we need a relative minus sign. If we have a positive quantity on the left-hand side the amount of copper increases and this must correspond to a net inflow.

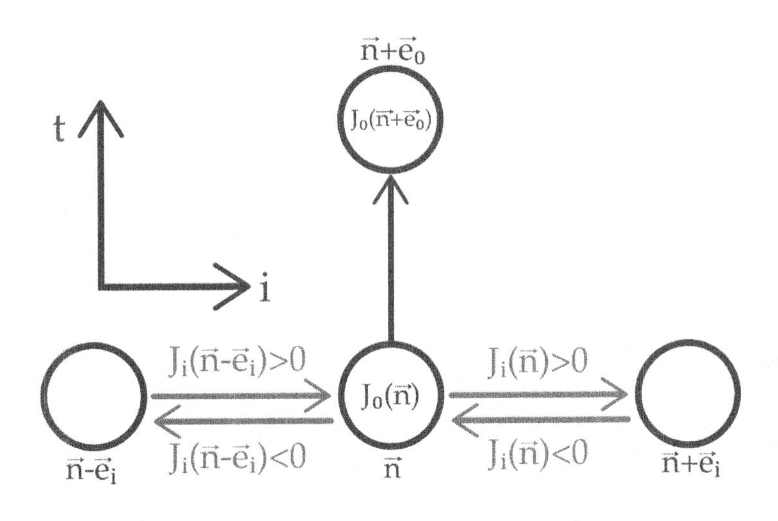

In the continuum limit, Eq. 2.31 becomes the famous **continuity equation**[19]

[19] Here and in the following, we use Einstein's summation convention, i.e. there is an implicit sum over all indices that appear in pairs. This allows us to write $\sum_{\mu=0}^{2} \partial_\mu J_\mu$ as $\partial_\mu J_\mu$.

$$\partial_0 J_0 + \sum_{i=1}^{2} \partial_i J_i = 0$$

$$\therefore \quad \sum_{\mu=0}^{2} \partial_\mu J_\mu = 0$$

$$\therefore \text{ summation convention} \quad \partial_\mu J_\mu = 0 \,. \tag{2.32}$$

Now, what we really want is an equation that lets us understand how arbitrage opportunities show up and evolve dynamically. Moreover, we already know that J_μ and $F_{\mu\nu}$ are good quantities to describe our system since these do not depend on local conventions. The quantities J_μ contain information about the positions and flow of copper, while $F_{\mu\nu}$ represent arbitrage opportunities.

However, the first naive guess to write $F_{\mu\nu}$ on one side and

J_μ on the other side of an equation fails because J_μ has only one index but $F_{\mu\nu}$ has two indices. In addition, there is one further piece of information that we can use: Eq. 2.32.

If taking the derivative of the right-hand side of an equation yields zero, the left-hand side has to be zero, too. There is, however, no reason why $\partial_\mu F_{\mu\nu}$ should be zero and this is another hint that our first naive guess is wrong.

The crucial observation is that

$$\partial_\nu \partial_\mu F_{\mu\nu} = 0\,, \tag{2.33}$$

since $F_{\mu\nu}$ is antisymmetric but partial derivatives commute $\partial_\mu \partial_\nu = \partial_\nu \partial_\mu$.[20] This suggests that we should try

$$\partial_\nu F_{\mu\nu} = \mu_0 J_\mu\,, \tag{2.34}$$

where we have introduced the proportionality constant μ_0 which encodes how strongly the pattern of arbitrage opportunities reacts to the presence and flow of copper.

This equation has exactly one free index (μ) on both sides and, most importantly, both sides yield zero if we calculate the derivative

$$\begin{aligned} \partial_\mu \partial_\nu F_{\mu\nu} &= \mu_0 \partial_\mu J_\mu \\ & \quad \text{Eq. 2.32 and Eq. 2.33} \\ 0 &= 0 \quad \checkmark \end{aligned}$$

Eq. 2.34 is the famous **inhomogeneous Maxwell equation**.

For our discrete system it reads

$$\sum_{\nu=0}^{2} F_{\mu\nu}(\vec{n}) - \sum_{\nu=0}^{2} F_{\mu\nu}(\vec{n} - \vec{e}_\nu) = \mu_0 J_\mu(\vec{n})\,. \tag{2.35}$$

[20] This is an important general result. Every time we have a sum over something symmetric in its indices multiplied with something antisymmetric in the same indices, the result is zero:

$$\sum_{ij} a_{ij} b_{ij} = 0$$

if $a_{ij} = -a_{ji}$ and $b_{ij} = b_{ji}$ holds for all i, j. We can see this by writing

$$\sum_{ij} a_{ij} b_{ij} = \frac{1}{2}\Big(\sum_{ij} a_{ij} b_{ij} + \sum_{ij} a_{ij} b_{ij}\Big)$$

We are free to rename our indices $i \to j$ and $j \to i$, which we use in the second term

$$\therefore \sum_{ij} a_{ij} b_{ij} = \frac{1}{2}\Big(\sum_{ij} a_{ij} b_{ij} + \sum_{ij} a_{ji} b_{ji}\Big)$$

Then we use the symmetry of b_{ij} and antisymmetry of a_{ij}, to switch the indices in the second term, which yields

$$\begin{aligned} \therefore \sum_{ij} a_{ij} b_{ij} &= \frac{1}{2}\Big(\sum_{ij} a_{ij} b_{ij} + \sum_{ij} \underbrace{a_{ji}}_{=-a_{ij}} \underbrace{b_{ji}}_{=b_{ij}}\Big) \\ &= \frac{1}{2}\Big(\sum_{ij} a_{ij} b_{ij} - \sum_{ij} a_{ij} b_{ij}\Big) \\ &= 0 \end{aligned}$$

Moreover, Eq. 2.35 with $\mu = 0$ reads

$$\sum_{\nu=0}^{2} F_{0\nu}(\vec{n}) - \sum_{\nu=0}^{2} F_{0\nu}(\vec{n} - \vec{e}_\nu) = \mu_0 J_0(\vec{n})$$

$\quad F_{00}(\vec{n}) = 0$

$$\sum_{i=1}^{2} F_{0i}(\vec{n}) - \sum_{i=1}^{2} F_{0i}(\vec{n} - \vec{e}_i) = \mu_0 J_0(\vec{n}), \tag{2.36}$$

where we used that $F_{00}(\vec{n}) = 0$, since $F_{\mu\nu}$ is antisymmetric.[21] Here, we have $\mu_0 J_0(\vec{n})$ on the right-hand side, which is proportional to the amount of copper located at $\vec{n}$. Thus, for $\mu = 0$, Eq. 2.34 gives us information about the pattern of exchange rates around a country in which copper is present. In the continuum limit, Eq. 2.36 becomes

$$\partial_i F_{0i} = \mu_0 J_0, \tag{2.37}$$

which is known as **Gauss's law**.

[21] An object is antisymmetric if we get a minus sign when we swap its indices: $F_{\mu\nu} = -F_{\nu\mu}$. This automatically implies that all components for which the two indices are equal will vanish. :

$$F_{00} = F_{11} = F_{22} = 0$$

because the antisymmetry implies $F_{00} = -F_{00}$, $F_{11} = -F_{11}$ and $F_{22} = -F_{22}$. This can only be true if $F_{00} = 0$, $F_{11} = 0$ and $F_{22} = 0$.

Similarly, for $\mu \to i \in \{1, 2\}$ in Eq. 2.35, we get equations that give us information about the pattern of exchange rates that are present whenever copper flows[22]

$$\sum_{\nu=0}^{2} F_{i\nu}(\vec{n}) - \sum_{\nu=0}^{2} F_{i\nu}(\vec{n} - \vec{e}_\nu) = \mu_0 J_i(\vec{n}). \tag{2.38}$$

In the continuum limit, this equation becomes[23]

$$\partial_0 F_{i0} + \partial_j F_{ij} = \mu_0 J_i, \tag{2.39}$$

which is known as the **Ampere-Maxwell law**.

[22] Above we investigated the temporal component of the equation $\mu = 0$. Now we investigate the spatial part $\mu \to i \in \{1, 2\}$.

[23] Again, Einstein's summation convention is used. Here this means that there is an implicit sum over the index j because it appears twice in the second term.

In addition to the inhomogeneous Maxwell equation (Eq. 2.34), there is the **homogeneous Maxwell equation**:

$$\partial_\lambda F_{\mu\nu} + \partial_\mu F_{\nu\lambda} + \partial_\nu F_{\lambda\mu} = 0. \tag{2.40}$$

This equations follows automatically from the definition of the gain factor $F_{\mu\nu}$ in terms of the exchange rates A_μ.

$$
\begin{aligned}
&\partial_\lambda F_{\mu\nu} + \partial_\mu F_{\nu\lambda} + \partial_\nu F_{\lambda\mu} \\
&= \partial_\lambda(\partial_\mu A_\nu - \partial_\nu A_\mu) + \partial_\mu(\partial_\nu A_\lambda - \partial_\lambda A_\nu) + \partial_\nu(\partial_\lambda A_\mu - \partial_\mu A_\lambda) && \text{Eq. 2.22} \\
&= \partial_\lambda\partial_\mu A_\nu - \partial_\lambda\partial_\nu A_\mu + \partial_\mu\partial_\nu A_\lambda - \partial_\mu\partial_\lambda A_\nu + \partial_\nu\partial_\lambda A_\mu - \partial_\nu\partial_\mu A_\lambda \\
&= \partial_\lambda\partial_\mu A_\nu - \partial_\lambda\partial_\nu A_\mu + \partial_\mu\partial_\nu A_\lambda - \partial_\lambda\partial_\mu A_\nu + \partial_\lambda\partial_\nu A_\mu - \partial_\mu\partial_\nu A_\lambda && \partial_\mu\partial_\nu = \partial_\nu\partial_\mu \\
&= \cancel{\partial_\lambda\partial_\mu A_\nu} - \cancel{\partial_\lambda\partial_\nu A_\mu} + \cancel{\partial_\mu\partial_\nu A_\lambda} - \cancel{\partial_\lambda\partial_\mu A_\nu} + \cancel{\partial_\lambda\partial_\nu A_\mu} - \cancel{\partial_\mu\partial_\nu A_\lambda} \\
&= 0 \quad \checkmark
\end{aligned}
$$

Now we move on and try to write down rules which describe how our traders behave.

2.2 Trader Dynamics

Before we try to understand how traders react to arbitrage opportunities, we need to understand how they behave when there are none.[24]

[24] In physical terms, we say that we first want to write down rules describing *free* traders, before we try to describe *interacting* traders.

But what do we actually know about traders? Of course, if there is a risk-less opportunity to make money, they will certainly use it. This is the easy part which we want to understand in a second step. But what do they do if there are no such obvious opportunities?

The best answer I can give you is: we don't know. The individual psychology of each trader is far too complex and it is simply impossible to know all possible factors which influence their decisions. But this doesn't mean that we can't describe how traders move through our system. In fact, many of the important advances in modern physics are methods to describe systems without knowing all the details.[25] And after all, it's one of the main goals of economists to produce theories of human behavior that gloss over the details of individual psychology.

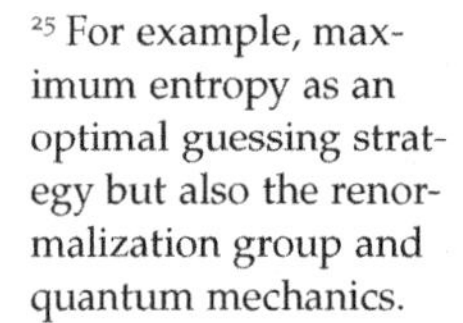

[25] For example, maximum entropy as an optimal guessing strategy but also the renormalization group and quantum mechanics.

On a "microscopic" level there are an incredibly large number of factors influencing the decisions of each trader all the time.

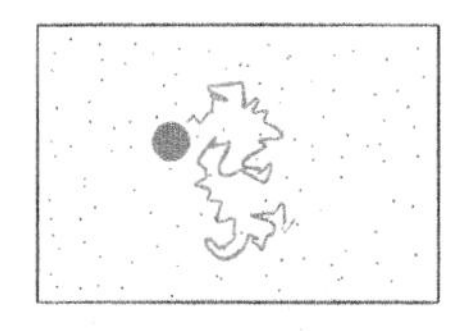

The situation is therefore analogous to what happens to pollen grains in water. These grains constantly collide with water molecules, but there is no way we can describe all the water molecules influencing the pollen grains. If we ignore this microscopic level and zoom out, our pollen grains simply perform random walks.

Therefore, our working assumption will be that traders

move randomly.

Of course, if we zoom in we can deduce how traders react to individual signals (Behavioral Psychology) analogous to how we can deduce how pollen grains collide with individual water molecules.

This doesn't really help us because there are too many factors influencing trader decisions and too many water molecules colliding with our pollen grains. From a macroscopic perspective, the random walk hypothesis is really the best we can do.

What we therefore need, is an equation describing our random walking traders.[26] Since we are not making any special assumptions, the equation we are looking for is exactly the same equation describing pollen grains in water.

This equation which describes random moving objects is known as the **diffusion equation** and reads[27]

$$\frac{\partial \rho(t,x)}{\partial t} = D\frac{\partial^2 \rho(t,x)}{\partial x^2}, \tag{2.41}$$

where ρ is the probability density of the quantity we want to describe and D is a constant known as diffusion constant.[28]

Previously, we argued that $J_0(t,x)$ tells us how much copper is located at a particular location.[29] But now, we are assuming that our traders move copper around randomly. This means that we need to describe our model in probabilistic terms.

To understand what this means, let's imagine that there exists just one unit of copper. Our goal is to describe how this unit of copper is moved around. In this case, we need to

[26] Maybe you're disappointed that we aren't trying something more ambitious. But be assured that we are steadily moving towards the edge of modern physics. One of the main realizations in modern physics was that the laws describing nature at a more fundamental level do not help you to describe nature if you zoom out. Quantum mechanics does not help us to describe how a ball rolls down a ramp because a ball consists of far too many quantum particles and there is no way to take the quantum rules governing their individual behavior into account. Analogously, there are too many actors and factors in an economy. Therefore, the rules governing the behavior of individual actors do not help us to describe the economy as a whole.

[27] For simplicity, we restrict ourselves to one spatial dimension. In three spatial dimensions, the diffusion equation reads

$$\frac{\partial \rho(t,\vec{x})}{\partial t} = D\nabla^2 \rho(t,\vec{x}),$$

where ∇^2 is the Laplacian operator.

[28] We will talk about these two new notions below.

[29] We discussed this in the text below Eq. 2.29.

describe it using a **probability density**.

For example, we can imagine that at one point in time t_0, the function $\rho(t, x)$ is exactly one for one particular location x_0:

$$\rho(t_0, x_0) = 1 \quad \text{and zero otherwise.} \tag{2.42}$$

In other words, this means that we are 100% certain to find our unit of copper at t_0 at the location x_0.

At a later point in time t_1, the probability density possibly looks like this:

$$\begin{aligned} \rho(t_1, x_0) &= 0.1 \\ \rho(t_1, x_1) &= 0.2 \\ \rho(t_1, x_2) &= 0.7 \\ \rho(t_1, x) &= 0 \quad \text{for all other locations } x \end{aligned} \tag{2.43}$$

This means that we are 70% certain to find the unit of copper at the location x_2, 20% certain to find it at x_1 and only 10% certain to find it at x_0. Formulated differently, if we prepare the system equally many times, we expect to find our unit of copper at this point in time at x_2 in 70% of cases, at x_1 in 20% of cases and at x_0 in only 10% of cases. Moreover, we don't expect to find it anywhere else. Similarly, if there are multiple units of copper in the system, $\rho(t, x)$ tells us how we expect them to be distributed throughout the system.

To summarize: J_0 previously told us how much copper is located at a particular location. Now, we use ρ to describe how much copper we can expect to find at a particular location if we repeat our experiment multiple times. This change to a probabilistic perspective is necessary because we assume that our traders move copper around randomly. While we therefore can't be certain what will happen, we can make probabilistic predictions.

Now, how do we get from the random-walk hypothesis to Eq. 2.41?[30]

[30] If you don't care about how we get from the random walk hypothesis to the diffusion equation, you can skip the rest of this subsection.

As before, let's focus on one unit of copper. At each time step Δt the unit's location x changes by a step l:

$$x(t+\Delta t) = x(t) + l(t) . \qquad (2.44)$$

Our task is to write down an equation which allows us to calculate the probability density $\rho(x, t+\Delta t)$ given the probability density at a previous time step $\rho(x', t)$. Then, we can start with a specific density at one point in time (for example one in which all copper is located at one particular point) and calculate how copper gets distributed through the system.[31] If a unit of copper moves from x' at time t to x at time $t+\Delta t$, it must have moved a step $l = x - x'$. What's the probability that it moves such a step?

[31] In more technical terms: given initial conditions, we can then calculate how the system evolves as time passes.

To answer this question, we need to introduce the **probability distribution** of our random walk.[32] We denote the probability distribution which tells us the probability for each possible step as $\chi(l)$. In particular, $\chi(l_s)$ tells us the probability that our unit of copper moves a specific step l_s.

[32] If you're unfamiliar with probability distributions, have a look at Appendix A.1.

The probability to find our unit of copper at x at time $t+\Delta t$ is therefore[33]

$$\begin{aligned}
\rho(x, t+\Delta t) &= \int_{-\infty}^{\infty} \rho(x', t)\chi(x - x')dx' \\
& \qquad \text{new variable } l \equiv x - x' \\
&= \int_{-\infty}^{\infty} \rho(x-l, t)\chi(l)dl , \qquad (2.45)
\end{aligned}$$

where we sum over all possible previous locations x'. In words, this formula means that we sum over all possibilities how our unit of copper can end up at x at $t+\Delta t$. In principle, we can reach x from any location x'. But we need to take into account that the step from x' to x only happens with a certain probability, as indicated by $\chi(x-x')$. Moreover, the probability density that indicates how likely is to

[33] Take note that the change of variable brings in a minus sign, but that this minus sign gets canceled by another minus sign that results from taking into account the limits of the integral, which get reversed for the new variable.

find our unit of copper at x' at t is $\rho(x',t)$. Hence, the probability to find the unit of copper at x at $t + \Delta t$ is given by the sum over all possible locations x' at t weighted by the probability density values that tell us how likely it is that it really was at x' and that the step from x' to x happens.

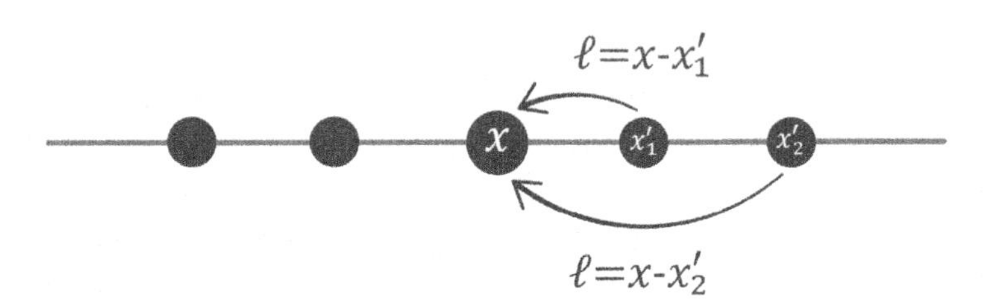

Now we need to make some (rather mild) assumptions about $\chi(l)$ in order to move forward. While we do not need to specify the probability distribution $\chi(l)$ itself, we need to make an assumption about the **mean** and **standard deviation** of the steps l.[34] From the mean and the standard deviation, we can tell how far from its original position we can expect to find each given unit of copper. But since we do not want to single out any particular direction, we assume the probability for steps in each direction are equal. This implies that the mean of l is zero because steps to the left happen on average equally often as steps to the right:[35]

[34] The notions of mean and standard deviation are discussed in Appendix A.

[35] We sum over the probability $\chi(l)$ for each step l.

$$\int_{-\infty}^{\infty} l\chi(l)dl = 0\,. \tag{2.46}$$

But, of course, we do not assume that copper doesn't get moved around at all. Instead a mean of zero only implies that steps in each direction happen equally likely and therefore average out. This also means that the standard deviation

$$\int_{-\infty}^{\infty} l^2\chi(l)dl = a \tag{2.47}$$

is non-zero.

The standard deviation encodes information about the total distance each unit of copper gets moved around *on average* at each step.[36] The constant a is some constant which depends on the system in question and usually needs to be determined experimentally. In other words, a is simply the name we give to the standard deviation. In addition, $\chi(l)$ has to be normalized properly[37]

$$\int_{-\infty}^{\infty} \chi(l)dl = 1 . \qquad (2.48)$$

[36] As mentioned above already, the meaning of the standard deviation is discussed in Appendix A.

[37] This is not an assumption but a consistency requirement.

This requirement ensures that the probabilities for all possible steps sum up to $100\% = 100\frac{1}{100} = 1$. With these formulas at hand, we can rewrite Eq. 2.45. The crucial idea is that we observe the system from far away and the stepsize is therefore small compared to the scale on which ρ varies noticeably.[38]

[38] We can imagine that our trades happen between neighboring individual locations while we are interested in the general behavior of the global market.

This allows us to Taylor expand $\rho(x-l,t)$:[39]

[39] The idea behind the Taylor expansion is explained in Appendix E.

$$\rho(x,t+\Delta t) = \int_{-\infty}^{\infty} \rho(x-l,t)\chi(l)dl$$

↓ Taylor expansion

$$\approx \int_{-\infty}^{\infty} \left(\rho(x,t) - l\frac{\partial \rho}{\partial x} + \frac{1}{2}l^2\frac{\partial^2 \rho}{\partial x^2}\right)\chi(l)dl$$

↓

$$\approx \rho(x,t) \underbrace{\int_{-\infty}^{\infty} \chi(l)dl}_{=1 \text{ (Eq. 2.48)}} - \frac{\partial \rho}{\partial x} \underbrace{\int_{-\infty}^{\infty} l\chi(l)dl}_{=0 \text{ (Eq. 2.46)}} + \frac{1}{2}\frac{\partial^2 \rho}{\partial x^2} \underbrace{\int_{-\infty}^{\infty} l^2\chi(l)dl}_{=a \text{ (Eq. 2.47)}}$$

↓

$$\approx \rho(x,t) + \frac{a}{2}\frac{\partial^2 \rho}{\partial x^2} . \qquad (2.49)$$

Moreover, we assume that the trades happen much quicker than the timescale at which we observe ρ. This allows us to

Taylor expand $\rho(x, t + \Delta t)$ on the left-hand side

$$\rho(x, t + \Delta t) \approx \rho(x, t) + \Delta t \frac{\partial \rho}{\partial t} . \quad (2.50)$$

Using Eq. 2.49 and Eq. 2.50 we therefore find

$$\rho(x, t) + \Delta t \frac{\partial \rho}{\partial t} \approx \rho(x, t) + \frac{a}{2} \frac{\partial^2 \rho}{\partial x^2}$$

$$\curvearrowright \; \cancel{\rho(x,t)}$$

$$\Delta t \frac{\partial \rho}{\partial t} \approx \frac{a}{2} \frac{\partial^2 \rho}{\partial x^2}$$

$$\curvearrowright$$

$$\frac{\partial \rho}{\partial t} \approx \frac{a}{2\Delta t} \frac{\partial^2 \rho}{\partial x^2}$$

$$\curvearrowright \; D \equiv \frac{a}{2\Delta t}$$

$$\frac{\partial \rho}{\partial t} \approx D \frac{\partial^2 \rho}{\partial x^2} , \quad (2.51)$$

where in the last step, we introduced the diffusion constant $D \equiv \frac{a}{2\Delta t}$. This is exactly the diffusion equation (Eq. 2.41) we wanted to derive.

We have therefore shown that the diffusion equation allows us to predict in probabilistic terms how copper gets moved around if our traders act randomly.

Now we can move on and think about how we can incorporate into our equation how traders react to arbitrage opportunities.

2.2.1 Trader Dynamics and Arbitrage

An arbitrage opportunity is a risk-less way to make money. So whenever such an opportunity shows up in the market it's clear that this will cause a drift among our traders.

How we can incorporate this effect into the diffusion equation?[40]

[40] So far the diffusion equation only describes how copper is moved through the system by traders in the absence of arbitrage opportunities

The answer to this question is really simple and beautiful. In fact, the idea we will use in the following is regarded by many physicists as one of the most fundamental insights in modern physics.

The main observation is that as soon as there are arbitrage opportunities, we need to be more careful whenever we calculate derivatives. When we calculate the derivative of a function, we compare the values of the function at two different locations. This follows if we recall the usual definition of a derivative in terms of a difference quotient[41]

$$\frac{\partial f(x)}{\partial x} \equiv \lim_{h \to 0} \frac{f(x+h) - f(x)}{h} . \qquad (2.52)$$

[41] Here we compare the value of the function $f(x)$ at the location x with its value at the location $x + h$.

But whenever there is an arbitrage opportunity, there is necessarily a non-trivial exchange rate A_μ somewhere.[42] Therefore, comparing the values of a function at two different locations is non-trivial. We need to take the exchange rate A_μ between the two locations into account. Otherwise we are comparing apples with oranges (e.g. dollars to euros).

[42] This follows because arbitrage opportunities are described by (Eq. 2.22):

$$F_{\mu\nu}(t, \vec{x}) \equiv \frac{\partial A_\nu}{\partial x^\mu} - \frac{\partial A_\mu}{\partial x^\nu} .$$

Therefore, an arbitrage opportunity exists if $F_{\mu\nu} \neq 0$ and this implies $A_\mu \neq 0$ for at least one of the components μ.

Luckily, we already know how to do this. The idea we need is exactly the same one that we used already in Eq. 2.27. Our goal there was to compare the prices of copper at two different locations and we found

$$J_\mu(\vec{n}) = q\Big(\varphi(\vec{n} + \vec{e}_\mu) - \varphi(\vec{n}) - A_\mu(\vec{n})\Big) , \qquad (2.53)$$

which in the continuum limit becomes

$$J_\mu(\vec{x}) = q\Big(\partial_\mu \varphi(\vec{x}) - A_\mu(\vec{x})\Big) , \qquad (2.54)$$

because the first two terms yield a difference quotient.

This is how, in general, the derivative of a function looks in the presence of non-trivial exchange rates $A_\mu(\vec{x})$! It is conventional to call this the **covariant derivative**

$$D_\mu \equiv \partial_\mu - A_\mu(\vec{x}) , \qquad (2.55)$$

In words, this means that we need to adjust the values of the function using the appropriate exchange rate.[43]

[43] The adjustment happens like this and not by multiplying the function with the exchange rate because we are working, as before, with the corresponding logarithms.

This is already everything we need to know.

To get the correct equation describing our traders in the presence of non-trivial exchange rates, we simply have to replace the ordinary derivative ∂_μ with the covariant derivative D_μ.[44]

[44] In physics, this procedure is known as minimal coupling.

Concretely, the diffusion equation in the presence of arbitrage opportunities reads[45]

[45] Reminder: the ordinary one-dimensional diffusion equation reads (Eq. 2.41)

$$\partial_t J_0(t,x) = D\partial_x^2 J_0(t,x)\,.$$

$$D_t J_0(t,x) = D D_x^2 J_0(t,x)\,,$$

$$\text{Eq. 2.55} \quad (\partial_t - A_t(t,x))J_0(t,x) = D(\partial_x - A_x(t,x))^2 J_0(t,x) \qquad (2.56)$$

where D is still the diffusion constant and D_x and D_t denote the covariant derivatives.[46]

[46] In physical terms, the new terms in the diffusion equation describe that there is a force which pushes our traders in a particular direction.

That's it. To describe how traders behave when arbitrage opportunities exist, we simply have to consistently take non-trivial exchange rates into account. We do this by acknowledging that whenever we calculate the derivative of some function, we need to include the appropriate exchange rate. This is necessary because when we calculate a derivative, we are evaluating the function at two different locations. This leads us to the conclusion that we need to replace our ordinary derivatives with covariant derivatives $\partial_\mu \to D_\mu$, where $D_\mu = \partial_\mu - A_\mu$. If we do this, we automatically end up with the correct equations.

There are two important lessons that we learned in this chapter. Firstly, we saw why gauge symmetry is useful. By thinking about gauge symmetry we learn which quantities are independent of local conventions. Secondly, we derived the correct equation describing the interplay between the flow of goods like copper and arbitrage opportunities by using the ideas that we should describe the system with gauge invariant quantities, that copper is conserved and that traders move randomly in the absence of risk-less opportunities to make money.

With this in mind, we are ready to understand how nature works at the fundamental level.

Part II
The Physics of Nature

"When learning about the laws of physics you find that there are a large number of complicated and detailed laws, laws of gravitation, of electricity and magnetism, nuclear interactions, and so on, but across the variety of these detailed laws there sweep great general principles which all the laws seem to follow."

Richard P. Feynman

Many of the most important discoveries in modern physics can be translated into the language of geometry and this allows us to understand them from a common perspective.

Specifically, one of the main discoveries of the last century was that our arena in physics consists of spacetime and three internal spaces known as charge space, isospin space and color space. This is important because there is a direct connection between each of these spaces and a specific fundamental interaction.

So far, four fundamental interactions have been discovered:[47]

[47] Sometimes the notion of fundamental forces is used instead of fundamental interactions. But this can lead to misunderstandings because the effects caused by fundamental interactions can be quite different from what we know from macroscopic forces. For example, weak interactions lead to the decay of nuclei and it is quite difficult to bring this into agreement with the macroscopic notion of a force.

- ▷ At large (cosmological) scales, gravity is the most important of the four interactions. Gravity is responsible for the movement of planets and galaxies.
- ▷ At tiny scales, strong and weak interactions are responsible for most of the interesting phenomena. For example, protons are held together by the strong force and nuclei decay through weak interactions.
- ▷ In between these two extremes it's electromagnetic interactions which hold sway.

While gravity is directly connected to spacetime, weak interactions are connected to isospin space, strong interactions to color space and electromagnetic interactions to charge space.

To understand all this, it makes sense to rip the total arena apart and study its individual constituents. In particular, we will discuss properties of the spacetime part and of the internal parts of the arena separately.

In addition, before we discuss global properties of the arena, it makes sense to study how it looks like locally. This is a useful approach because, while the arena as a whole is extremely complicated, locally its structure is quite simple.[48]

[48] This is analogous to what we discussed for the financial market previously. The internal space above one country is simply a line. But globally these lines can be glued together non-trivially and therefore, in general, our arena as a whole is quite complicated.

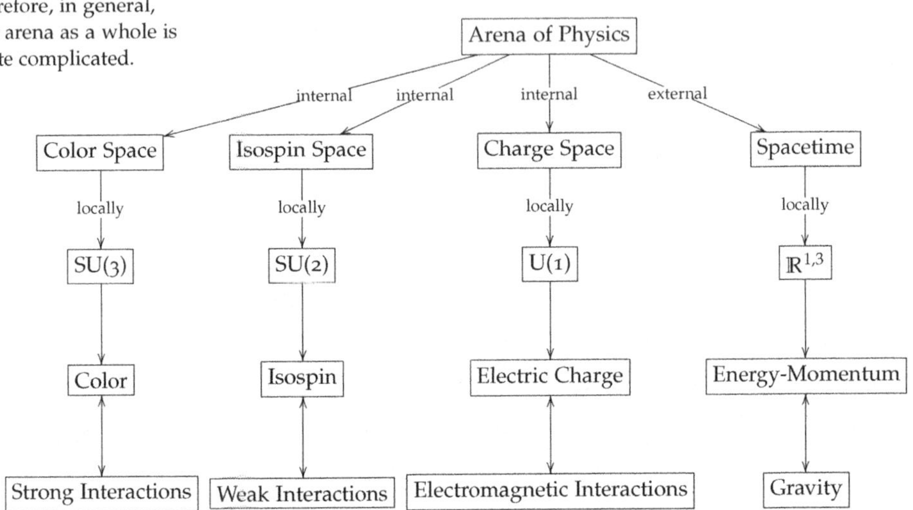

Understanding all this may sound like a lot of work. But the good news is that everything that we will discuss in the following sections is completely analogous to what we discussed in the previous chapters. In other words, if you've understood the main ideas discussed in the previous chapters, you are perfectly prepared to understand how nature works at the fundamental level. Moreover, the general discussion for each of the individual parts is extremely similar; only the details are different.

But before we discuss the properties of the arena, we should talk about the objects we want to describe.

3

Elementary Particles

During the last century, physicists discovered a whole zoo of elementary particles. For all we presently know, these particles cannot be further decomposed and have no substructure. That's why we call them elementary particles.

These elementary particles are the building blocks that all of the more complex structures (atoms, molecules, proteins) we see around us consist of. For example, a hydrogen atom consists of a proton and an electron.

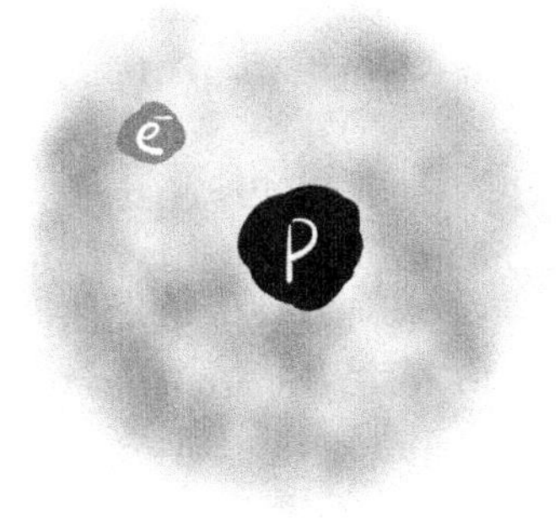

But the proton is not an elementary particle. In fact, it consists of quarks.[1]

[1] To be a bit more precise: a proton consists of two up quarks and one down quark.

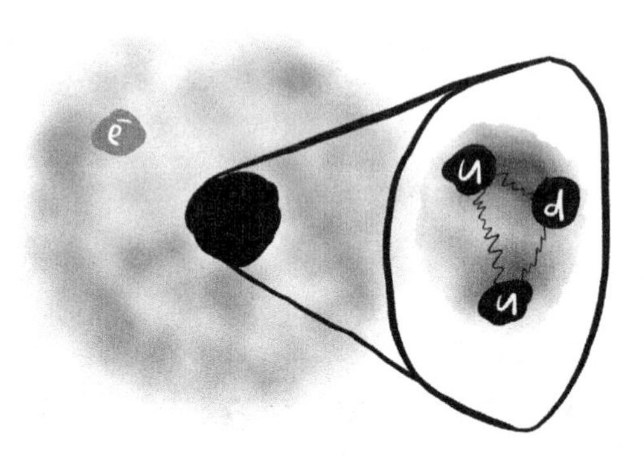

Quarks and electrons are, as far as we know, elementary particles.[2]

[2] Take note that there are lots of additional elementary particles which cannot be discovered so easily (muons μ, tauons τ, charm quarks, strange quarks, top quarks, bottom quarks). They are heavy cousins of the electron and the quarks that make up a proton (up quark and down quark). Since they are heavier, they decay into their lighter cousins shortly after they are produced. They were discovered in collider experiments in which high-energetic particles collide. When two protons collide at higher energies, they sometimes create heavier particles. (We will understand roughly how this works below.) In this sense, colliders are essentially large microscopes which allow us to look deeper at matter. The higher the energy of the colliding particles, the deeper we can look.

The particle zoo itself is not that interesting. Instead, what is interesting is the way the particles interact with each other.

In the toy model discussed in the previous chapter, we wanted to describe traders who carry copper around. Now our goal is to describe elementary particles which carry what we call charges around.[3]

[3] Take note that charge does not necessarily mean electric charge.

We can distinguish between different types of elementary particles because they carry different charges. If two elementary particles carry the same charges, we say that they are the same type of elementary particle, e.g., electrons.

To understand what we mean by charges, we need to recall one main idea. One of the deepest insights during the last century was that the arena of physics is not a static background structure. Instead, the arena itself is a dynamical actor which changes all the time. Most importantly, the arena is directly influenced by the behavior of all elementary particles and, in turn, these particles are influenced by the structure of the arena.

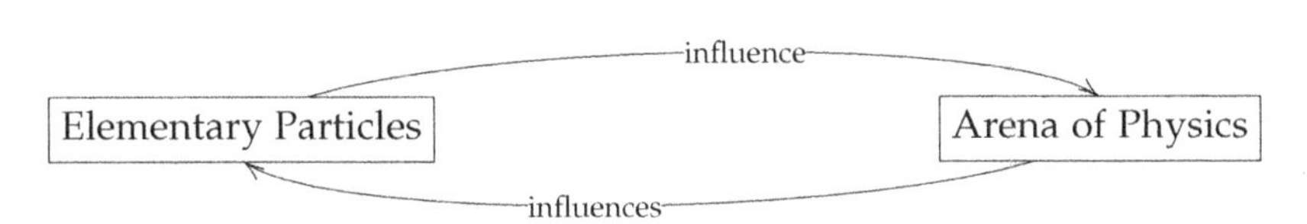

Formulated a bit more precisely, the way elementary particles influence each other is by modifying the geometry of the arena.

For example, if two elementary particles influence each other electromagnetically, this happens because they change the structure of the internal charge space and at the same time feel the structure of this charge space.[4]

But the crucial point is that elementary particles influence and get influenced by different parts of the arena. This is what we use to label elementary particles.

In other words, we identify elementary particles based on which part(s) of the total arena they modify and how strong their effect is on the corresponding geometries.[5]

Concretely:[6]

- ▷ We say that particles which modify charge space carry **electric charge**.
- ▷ We say that particles which modify isospin space carry **isospin**.
- ▷ We say that particles which modify color space carry **color**. (Take note that there is a reasonable explanation for the somewhat strange label "color" but it has nothing

[4] This is analogous to what we discussed in the previous chapter. There our traders are influenced by the geometry of the internal money space and, in turn, the geometry of the money space changes dynamically depending on the behavior of the traders.

[5] We can imagine that a trader who carries a lot of copper around has a much bigger effect than a trader who carries only a small amount.

[6] There is a beautiful connection between these labels and the symmetries of internal spaces. This connection is known as Noether's theorem and we will talk a bit more about it in Section 5.3. Moreover, take note that there is another incredibly important label called **spin**. While energy-momentum is the Noether charge following from invariance under spacetime translations, spin is the Noether charge following from invariance under rotations.

to do with the colors we perceive in everyday life. We will talk more about this when we discuss color space below.)

▷ We say that particles which modify spacetime carry **energy-momentum**. However, it is conventional to use the label "mass" for elementary particles because energy-momentum depends on the standpoint (reference frame) from which we observe the particle. (Mass is directly related to the "length" of the energy-momentum vector; and it is the quantity that is invariant (same in all reference frames). We will discuss this further in Section 5.3.)

With this in mind, here is a table listing a few of the elementary particles whose interactions we want to describe:

	electric charge	isospin	color	mass
left-chiral electron-neutrino ν_e	0	1/2	0	???
left-chiral electron e_L	$-e$	-1/2	0	m_e
right-chiral electron e_R	$-e$	0	0	m_e
red left-chiral up quark u_L^r	2/3 e	1/2	$(\frac{1}{2}, \frac{1}{2\sqrt{3}})$	m_u
blue left-chiral up quark u_L^b	2/3 e	1/2	$(0, \frac{-1}{\sqrt{3}})$	m_u
green left-chiral up quark u_L^g	2/3 e	1/2	$(\frac{-1}{2}, \frac{1}{2\sqrt{3}})$	m_u
red left-chiral down quark d_L^r	-1/3 e	-1/2	$(\frac{1}{2}, \frac{1}{2\sqrt{3}})$	m_d
blue left-chiral down quark d_L^b	-1/3 e	-1/2	$(0, \frac{-1}{\sqrt{3}})$	m_d
green left-chiral down quark d_L^g	-1/3 e	-1/2	$(\frac{-1}{2}, \frac{1}{2\sqrt{3}})$	m_d
red right-chiral up quark u_R^r	2/3 e	0	$(\frac{1}{2}, \frac{1}{2\sqrt{3}})$	m_u
blue right-chiral up quark u_R^b	2/3 e	0	$(0, \frac{-1}{\sqrt{3}})$	m_u
green right-chiral up quark u_R^g	2/3 e	0	$(\frac{-1}{2}, \frac{1}{2\sqrt{3}})$	m_u
red right-chiral down quark d_R^r	-1/3 e	0	$(\frac{1}{2}, \frac{1}{2\sqrt{3}})$	m_d
blue right-chiral down quark d_R^b	-1/3 e	0	$(0, \frac{-1}{\sqrt{3}})$	m_d
green right-chiral down quark d_R^g	-1/3 e	0	$(\frac{-1}{2}, \frac{1}{2\sqrt{3}})$	m_d

Here $e = 1.602 \times 10^{-19}$ C is the magnitude of the electric charge of an electron and $m_e = 9.109 \times 10^{-31}$ kg is its mass. The masses of neutrinos have not been measured and right-chiral neutrinos have not been discovered yet.

The remaining values for the masses are not listed because they are not important for our purpose.[7] In fact, the specific values of the charges of the various elementary particles are not important for the modest goal of this book.

The only message to take away is that there are different particles which carry different charges.

Take note that there are other elementary particles. There are two additional generations which consist of heavy cousins of the particles listed here. In other words, these additional generations of elementary particles have exactly the same charges, except for a larger mass. For example, the muon has exactly the same charges as the electron but a larger mass. Moreover, all the particles listed in the table, together with their heavier cousins of the other two generations, share a common property, as a result of which they are named, collectively, fermions. This property is that no two identical fermions can locally be found in the same quantum state.

In addition, other particles, collectively named bosons, share a different property, in that an indefinite number of them can be found in the same quantum state. Among these are the Higgs boson, and so-called gauge bosons called photons, W-bosons, Z-Bosons, gluons (and maybe gravitons). But the role of these particles in nature is a different one. While fermions are associated with matter, gauge bosons are associated with the fundamental interactions.[8]

In case you're wondering: no one knows why exactly these elementary particles exist and why they carry the labels

[7] You can find them, for example, on Wikipedia. Moreover, take note that the labels "left-chiral" and "right-chiral" refer to another important charge particles can carry known as chirality. But unfortunately, discussing chirality would lead us too far astray. For our purposes, it is sufficient to realize that the right-chiral electron is a different particle since it carries different charges. "Right-chiral electron" is simply the name we give to this particle. We will talk about the closely related concept of spin in Section 5.3. You can find a complete discussion in

Jakob Schwichtenberg. *Physics from Symmetry*. Springer, Cham, Switzerland, 2018b. ISBN 978-3319666303

[8] Formulated more technically, the particles listed in the table all carry spin $1/2$. (A fermion is defined as a particle with half-integer spin $1/2, 3/2, 5/2$ etc.) Particles with spin 1 are known as gauge bosons and they appear as quantized excitations of our bookkeepers A_μ, W_μ, G_μ which we will talk about below. (A boson is defined as a particle with integer spin $0, 1, 2$ etc.) In quantum field theory, we therefore say that these gauge bosons *mediate* their corresponding interaction.

they do. This is one of the big open questions in modern physics. We only know that these particles exist because we can observe them in experiments.

No one predicted their existence on theoretical grounds. For example, when the muon (a heavy cousin of the electron) was discovered, Nobel laureate I. I. Rabi famously quipped, "Who ordered that?"

Nevertheless, we can successfully describe the behavior and interactions of the various members of this particle zoo. We will do this again in two steps. First, we will determine how our particles behave when they are alone and only afterwards will we discuss how particles interact with each other. [9]

[9] PS: If you want to learn more about color charge, try `http://jakobschwichtenberg.com/what-are-quantum-numbers/`.

4 Quantum Mechanics

In Section 2.2, we discussed how we can describe the movement of copper through our system even though we can't take all factors influencing trader decisions into account. The main idea was that if we don't know all these details, we have to assume that copper gets moved around randomly.[1]

Using this assumption we were able to derive the so-called diffusion equation (Eq. 2.41)

$$\frac{\partial \rho(t,\vec{x})}{\partial t} = D\nabla^2 \rho(t,\vec{x}) . \tag{4.1}$$

Now we want to describe elementary particles.

Luckily, the situation is not so different. It was one of the biggest discoveries in modern physics that there seems to be a fundamental randomness in the behavior of elementary particles.

The origin and meaning of this randomness is something

[1] Recall that this is a reasonable assumption as long as there are no arbitrage opportunities. If there are risk-less opportunities to earn money, it's clear that our traders will use them to make money. But we only included this in a second step once we have understood how traders behave when there are no such opportunities.

that physicists have now argued about for more than a century and is not what I want to dive into here.[2]

[2] Just for the record: deriving the laws of quantum mechanics from the assumption that elementary particles perform random walks is known as the stochastic interpretation. One way to explain the randomness is by noting that in quantum field theory the vacuum (ground state) is not trivial. Instead, for each quantum field there are **vacuum fluctuations** which cannot be turned off. In other words, even in the state with the lowest possible energy, quantum fields still wiggle a bit around. Another popular way of looking at it is to say that virtual particles continuously pop in and out of existence. With this picture in mind, we can imagine that our elementary particles scatter continuously with these vacuum fluctuations/ virtual particles. In this sense, the situation is somewhat analogous to pollen grains in water which collide with the water molecules all the time.

The equation at the heart of quantum mechanics is known the **Schrödinger equation** and reads

$$\frac{\partial \psi(t,\vec{x})}{\partial t} = iD\nabla^2 \psi(t,\vec{x}) \,. \tag{4.2}$$

The only difference between Schrödinger's equation and the diffusion equation is the imaginary unit i on the right-hand side. In other words, Schrödinger's equation is a diffusion equation with an imaginary diffusion constant.

This small change has huge implications.

For example, while solutions of the diffusion equation $\rho(t,x)$ are real, solutions of the Schrödinger equation $\psi(t,\vec{x})$ are, in general, complex functions. This follows automatically because, to get the same expression on both sides of the equation, we must get an imaginary unit from somewhere. Specifically, on the left-hand side, $\frac{\partial \psi(t,\vec{x})}{\partial t}$ must yield something times i.

We can interpret any real function in a straightforward manner, but for a complex function the interpretation is less direct. It is clear that the result of any physical measurement is a real number; together with appropriate units. Thus, the complex function $\psi(t,\vec{x})$ itself cannot be directly related to a real physical quantity; its relationship to things we can measure has therefore to be somewhat indirect.

In contrast, we are able to interpret $\rho(t,\vec{x})$ directly as the probability density which tells us how much copper we can expect to find at a particular location.

So what's the meaning of $\psi(t,\vec{x})$?

While $\psi(t,\vec{x})$ is, in general, complex, it's quite easy to use it to construct a function which is certainly always real:[3]

$$\rho(t,\vec{x}) = \psi^\star(t,\vec{x})\psi(t,\vec{x})\,. \tag{4.3}$$

[3] A complex number multiplied by its complex conjugate always yields a real number. To see this, we note that we can write a complex number as a sum of its real part plus i times its imaginary part

$$z = a + ib$$

where a is the real part of z, b is its imaginary part and i is the imaginary unit which is defined by its property $i^2 = -1$. Using this notation we can calculate

$$\begin{aligned} &z^\star z = \\ &(a+ib)^\star(a+ib) = \\ &(a-ib)(a+ib) = \\ &a^2 + iab - iab - i^2b^2 = \\ &a^2 + b^2 \end{aligned}$$

(Take note how the imaginary unit i drops out and we end up with a purely real number.)

The superscript star symbol, $\star$, on a function denotes complex conjugation; that is, it denotes the *same* initial function *except* for the imaginary unit i replaced everywhere by $-i$ it appears in the initial function. The function $\rho(t,\vec{x})$ is interpreted as the **probability density in quantum mechanics**, analogous to how we interpreted $\rho(t,\vec{x})$ in the financial toy model.

We can understand how this interpretation comes about by noting that there is also a continuity equation in quantum mechanics which can be derived directly from the Schrödinger equation:[4]

$$\partial_t(\psi^\star\psi) = \frac{D}{i}\partial_i\Big(\psi\partial_i\psi^\star - \psi^\star\partial_i\psi\Big)\,. \tag{4.4}$$

[4] We will derive this explicitly below.

This equation has exactly the same structure as the continuity equation (Eq. 2.32). Any equation of this form is called a continuity equation. We get an equation of this form whenever there is a conserved quantity.

We therefore write it as

$$\partial_t\rho = \partial_i j_i\,. \tag{4.5}$$

This is why we interpret $\rho \equiv \psi^\star\psi$ as the probability density.[5]

[5] The quantity

$$j_i \equiv \frac{D}{i}\Big(\psi\partial_i\psi^\star - \psi^\star\partial_i\psi\Big)$$

is known as the **probability current**.

To make the connection to our previous continuity equation even more explicit, we rewrite Eq. 4.5 as

$$\partial_0 j_0 = \partial_i j_i.$$

Now, where does Eq. 4.4 come from?[6]

[6] As usual, if you're not interested in the derivation, feel free to skip the following paragraph.

One puzzle piece is to multiply the Schrödinger equation (Eq. 4.2) with the complex conjugated function $\psi^\star$

$$\psi^\star(t,\vec{x})\frac{\partial\psi(t,\vec{x})}{\partial t} = iD\psi^\star(t,\vec{x})\nabla^2\psi(t,\vec{x})\,. \tag{4.6}$$

The second puzzle piece we need is the complex conjugate of the Schrödinger equation (Eq. 4.2)

$$\frac{\partial\psi^\star(t,\vec{x})}{\partial t} = -iD\nabla^2\psi^\star(t,\vec{x})\,, \tag{4.7}$$

which we multiply with ψ:

$$\psi(t,\vec{x})\frac{\partial\psi^\star(t,\vec{x})}{\partial t} = -iD\psi(t,\vec{x})\nabla^2\psi^\star(t,\vec{x})\,. \tag{4.8}$$

The sum of Eq. 4.6 and Eq. 4.8 reads[7]

[7] To unclutter the notation, from now on we write ψ instead of $\psi(t,\vec{x})$ etc.

$$\psi^\star\frac{\partial\psi}{\partial t} + \psi\frac{\partial\psi^\star}{\partial t} = iD\psi^\star\nabla^2\psi - iD\psi\nabla^2\psi^\star\,. \tag{4.9}$$

We can write this more compactly by noting that[8]

[8] Hint: product rule.

$$\psi^\star\frac{\partial\psi}{\partial t} + \psi\frac{\partial\psi^\star}{\partial t} = \frac{\partial}{\partial t}(\psi^\star\psi)\,. \tag{4.10}$$

In addition, we note that

$$\psi^\star\nabla^2\psi - \psi\nabla^2\psi^\star = \nabla\left(\psi^\star\nabla\psi - \psi\nabla\psi^\star\right) \tag{4.11}$$

Therefore, we can write Eq. 4.9 as

$$\frac{\partial}{\partial t}(\psi^\star\psi) = iD\nabla\left(\psi^\star\nabla\psi - \psi\nabla\psi^\star\right)\,. \tag{4.12}$$

This is exactly the continuity equation (Eq. 4.4).[9]

[9] Take note that $\frac{1}{i} = \frac{-i^2}{i} = -i\frac{i}{i} = -i$

In summary, we have learned that the fundamental equation of quantum mechanics is extremely similar to the diffusion equation. The only difference between the two is that there is an imaginary unit i in the Schrödinger equation. It is really this little difference which makes quantum mechanics a bit more difficult than our financial toy model. In particular, we can't interpret solutions of the Schrödinger equation directly. Instead, we end up with the interpretation that the absolute square of the solutions yield probability densities.

Maybe you wonder why macroscopic objects behave so regularly if the elementary particles they consist of move around randomly?

This is possible because, while individual elementary particles seem to move around randomly, we can predict how they behave on average. Otherwise we couldn't write down equations like the diffusion equation and Schrödinger's equation.

Moreover, take note that macroscopic objects consist of an incredibly large number of elementary particles. The random behavior of the individual constituents therefore averages out and this allows us to describe macroscopic objects using simple laws like Newton's second law.[10]

[10] In quantum mechanics, this can be shown explicitly. This result is known as the **Ehrenfest theorem**.

Of course, there is a lot more to quantum mechanics than just the Schrödinger equation and its solutions. But, discussing quantum mechanics in detail requires at least a few hundred pages.[11] For the modest goal of this book we already have everything we need to move forward.

[11] If you want to learn more about quantum mechanics, try

Jakob Schwichtenberg. *No-Nonsense Quantum Mechanics*. No-Nonsense Books, 2018c. ISBN 978-1719838719

4.1 Charge Space

One of the big surprises in modern physics was that the internal part of the arena of physics is much bigger and more complex than anyone would have expected. While charge space is relatively simple and was discovered more than a century ago, isospin space and color space are more complicated and were discovered much later.

But the good news is that we can understand all of the essential features by focusing on the internal charge space. Once we've discussed charge space thoroughly, we can understand the structure of the more complicated color and isospin spaces almost immediately.

To understand charge space, we need to recall a few things which we have discussed already in the context of our financial toy model.

In this toy model, we discovered an internal space "above" each possible location. Each such internal space is simply a line going from $0.0\ldots01$ to ∞. Each point on this line represents a possible price of copper. The total internal space is given by all these lines taken together. We also discussed that when countries use independent local currencies, we need something which glues the various lines together. In our toy model, the exchange rates were the glue. Moreover, we argued that there is a symmetry because each country can rescale its currency freely. This is possible because as soon as a country rescales its currency, the exchange rates are adjusted immediately and there is no effect on the dynamics of the system. But we also discovered that the structure of the internal space can have a direct impact on the dynamics. This happens whenever there are imperfections in the exchange rates such that there are currency arbitrage opportunities. Whenever such a risk-free opportunity to earn money exists, traders will use it to earn money. Therefore, the dynamics of the system is crucially shaped by such arbitrage opportunities.

Now, in order to understand charge space and thus how elementary particles influence each other via electromagnetic interactions, we only need to repeat the same story using different actors and a slightly different internal space.

Let's start with the internal space. The existence of the internal space we call charge space follows directly from one of the basic facts of quantum mechanics.[12]

[12] Don't worry if you don't know any quantum mechanics. The following discussion is aimed at people without previous knowledge of quantum mechanics. Of course, we can't discuss quantum mechanics properly here. But if you want to learn more about quantum mechanics, you might enjoy reading

Jakob Schwichtenberg. *No-Nonsense Quantum Mechanics*. No-Nonsense Books, 2018c. ISBN 978-1719838719

In quantum mechanics, we describe what is going on using a *complex* function $\psi(t, \vec{x})$, called the wave function. However, everything we can measure is a *real* number. This means that the relationship between quantities, O, we can measure and the wave function ψ is always of the form $O \simeq \psi^\star \psi$. A direct consequence of this observation is that we have some freedom in our wave functions, analogous to the freedom we discovered in the price of copper. In particular, we always have the freedom to multiply our wave functions by a **phase factor**

$$\psi \to \tilde{\psi} = e^{i\epsilon}\psi, \tag{4.13}$$

since

$$\begin{aligned} O \simeq \psi^\star \psi \quad \Rightarrow \quad \tilde{\psi}^\star \tilde{\psi} &= (e^{i\epsilon}\psi)^\star e^{i\epsilon}\psi \\ & \quad \text{↷ } (e^{ix})^\star = e^{-ix} \\ &= e^{-i\epsilon}\psi^\star e^{i\epsilon}\psi \\ & \quad \text{↷} \\ &= \psi^\star e^{-i\epsilon} e^{i\epsilon}\psi \\ & \quad \text{↷ } e^{-i\epsilon}e^{i\epsilon} = 1 \\ &= \psi^\star \psi = O \quad \checkmark \end{aligned} \tag{4.14}$$

What we've discovered here is completely analogous to what we discussed for the financial toy model. In the financial toy model, we used the prices to describe copper and discovered that we have the freedom to rescale the prices by multiplying them by a real number (Eq. 2.2)

$$p \to e^{\epsilon} p. \tag{4.15}$$

In quantum mechanics, we describe particles, say an electron, using a wave function and have the freedom to multiply it by a phase factor (Eq. 4.13). The only difference is the little i in the exponential function.

This observation allows us to understand what the internal space of quantum mechanics is. In the financial toy model, the internal space was a line ranging from $0.0\ldots01$ to ∞ which represents all possible prices of copper. We can imagine that we determine the structure of this space by starting with one specific price and then applying all allowed symmetry transformations. This yields the space of all allowed values.

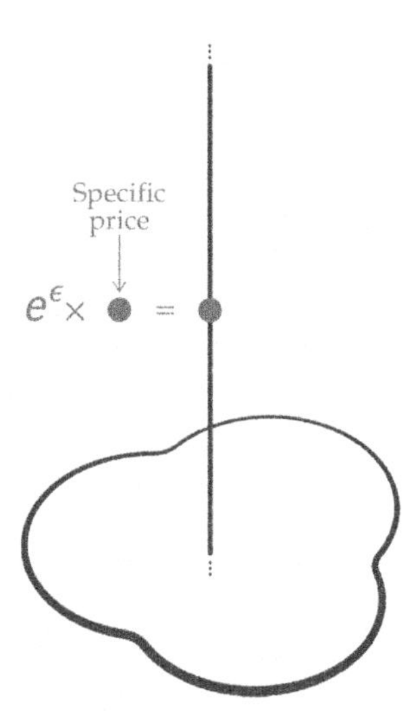

To get the internal space of quantum mechanics, we do exactly the same thing. We imagine that someone hands us a specific wave function Ψ:[13]

[13] A wave function is a complex function which can always be written in this form.

$$\psi(t,\vec{x}) = R(t,\vec{x})e^{i\varphi(t,\vec{x})}\,, \tag{4.16}$$

where $R(t,\vec{x})$ is a real function and $e^{i\varphi(t,\vec{x})}$ is a specific phase factor. We can then determine the structure of the corresponding internal space by applying all allowed symmetry transformations (Eq. 4.13). A crucial observation is now that this does not yield a line ranging from $0.0\ldots01$ to ∞ but a circle! Something of the form $e^{i\varphi(t,\vec{x})}$ evaluated at a specific location $\vec{x}$ and at a specific moment in time t is a complex number of unit length since

$$|e^{i\varphi(t,\vec{x})}|^2 = (e^{i\varphi(t,\vec{x})})^\star e^{i\varphi(t,\vec{x})} = e^{-i\varphi(t,\vec{x})} e^{i\varphi(t,\vec{x})} = 1 . \quad (4.17)$$

Geometrically all complex numbers of length one lie on a circle with radius one in the complex plane.

We can understand this using Euler's formula

$$\begin{aligned} z &= e^{i\phi} \\ &= \cos(\phi) + i\sin(\phi) \\ &\equiv \text{Re}(z) + i\text{Im}(z). \end{aligned}$$

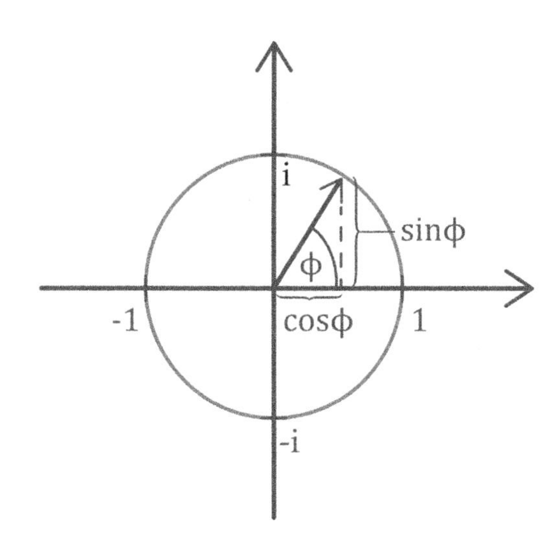

What this means is that in quantum mechanics, we have a little circle "above" each possible location.

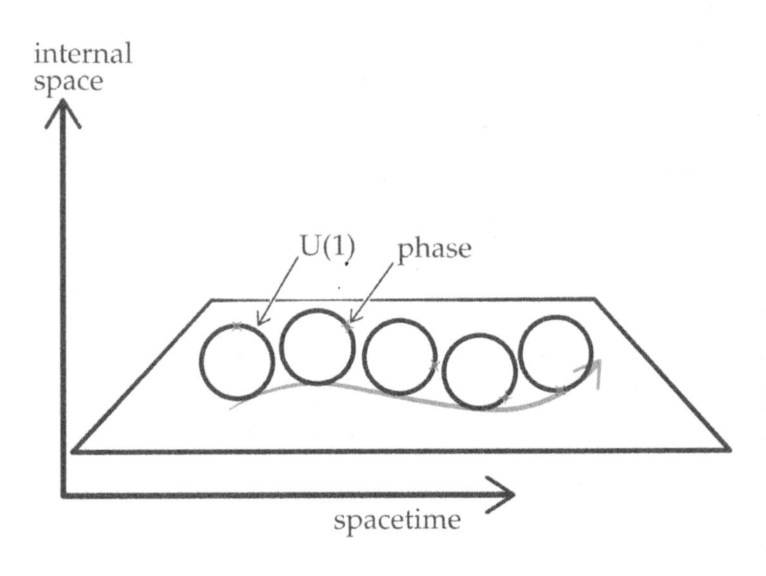

Each point on these circles represents a possible phase factor. Completely analogously to what we discussed for the prices of copper, these phase factors do not have any direct meaning because we can shift them however we want.[14] However, analogous to how the prices are, after all, essential to describe copper, the phase factors are essential to describe elementary particles. The main observation is again that the relationship is a bit more indirect than what we would've naively thought.

[14] This is discussed in a bit more detail in Appendix C.2.

That's really the only new thing we needed to understand. All that is left to do to derive a theory describing electromagnetic interactions is to apply all the lessons we have learned in the financial toy model to our new internal (charge) space.

4.2 Electrodynamics

First of all, we can imagine that we want to allow for the phase factor to be independently shifted at each location.

Of course, this is again possible but we then need something which keeps track of the local shifts. In other words, if we allow the phase factor to be shifted at each location independently, we need something which glues the circles together in such a way that the dynamics of the system remains unaffected by such local shifts. This is analogous to the exchange rates in our financial toy model which get adjusted every time a local currency is changed. Again, we denote the glue by A_μ. But, in physics, it is conventional to call A_μ the **electromagnetic potential**.

The most interesting part of the story is once more how the structure of the internal space affects the dynamics within the system. And again, we only need to recall what we have learned in the financial toy model. There we argued that currency arbitrage opportunities have a nontrivial effect on the dynamics. These are described by the object (Eq. 2.22)

$$F_{\mu\nu}(t,\vec{x}) \equiv \frac{\partial A_\nu(t,\vec{x})}{\partial x^\mu} - \frac{\partial A_\mu(t,\vec{x})}{\partial x^\nu} \tag{4.18}$$

and arise whenever there is copper present in the system.[15]

Completely analogously, we now introduce

$$F_{\mu\nu}(t,\vec{x}) \equiv \frac{\partial A_\nu(t,\vec{x})}{\partial x^\mu} - \frac{\partial A_\mu(t,\vec{x})}{\partial x^\nu} \tag{4.19}$$

for the electromagnetic potential A_μ.

In this context, we call $F_{\mu\nu}$ the **electromagnetic field tensor** and its components are usually called the **electric field** and the **magnetic field**.[16]

[15] The interplay between copper and arbitrage opportunities is described by (Eq. 2.34)

$$\partial_\nu F_{\mu\nu} = \mu_0 J_\mu \,.$$

[16] In other words, it is conventional to give special names to certain components of the electromagnetic field tensor $F_{\mu\nu}$. Since, $\mu,\nu \in \{0,1,2,3\}$ the field tensor can be written as a (4×4) matrix:

$$F_{\mu\nu} = \begin{pmatrix} F_{00} & F_{01} & F_{02} & F_{03} \\ F_{10} & F_{11} & F_{12} & F_{13} \\ F_{20} & F_{21} & F_{22} & F_{23} \\ F_{30} & F_{31} & F_{32} & F_{33} \end{pmatrix} = \begin{pmatrix} 0 & -E_1/c & -E_2/c & -E_3/c \\ E_1/c & 0 & -B_3 & B_2 \\ E_2/c & B_3 & 0 & -B_1 \\ E_3/c & -B_2 & B_1 & 0 \end{pmatrix}$$

We can write the field tensor like this, because it is antisymmetric: $F_{\mu\nu} = -F_{\nu\mu}$. This follows automatically from the definition in terms of the potential A_μ. The antisymmetry implies that there must be zeroes on the diagonal since, for example, $F_{11} = -F_{11}$, which is only true if $F_{11} = 0$. Moreover, we have, for example, $F_{12} = -F_{21}$ and therefore there are in total only 6 independent components.

Moreover, the equation (Eq. 2.34)

$$\partial_\nu F_{\mu\nu} = \mu_0 J_\mu \tag{4.20}$$

holds again. But this time, J_μ does not describe the flow of copper but the flow of electric charges through the system. In other words, this equation describes the interplay between electric charges and the electromagnetic field.

That's it. We've successfully derived electrodynamics.

Admittedly, we have glossed over lots of details. Nevertheless, it's true that we've successfully derived the correct structure describing the interplay between electric charges and the electromagnetic field solely by applying the lessons we learned in the financial toy model.[17]

[17] If you want to understand electrodynamics in more detail, you might enjoy my book

Jakob Schwichtenberg. *No-Nonsense Electrodynamics*. No-Nonsense Books, 2018a. ISBN 978-1790842117

Let's summarize what is going on here.[18]

[18] Don't worry if you're confused. We will discuss everything we just discovered multiple times from various perspectives.

In quantum mechanics, we use complex functions although everything we can measure in experiments has real values. This means that we can multiply the functions we use in quantum mechanics to describe what is going on with (unit) phase factors without changing anything.

This gauge symmetry is completely analogous to the gauge symmetry we discussed in the financial toy model. We therefore concluded that we also have an internal space, analogous to the money space in the toy model. The only difference is that now the internal space above each point is a circle and not a line. Again, the total internal space is all these local internal spaces taken together.

The main idea is then that this internal space is essential to describe the interactions of elementary particles analogous to how the internal money space is essential to describe how copper gets moved around in our toy model. In particular, there are elementary particles which have a direct influence on the structure of the internal space. We call these particles electrically charged particles. Wherever a charged particle is present, the structure of the internal space gets modified such that the various circles are glued together non-trivially.[19]

[19] This is analogous to how currency arbitrage opportunities show up whenever exchange rates favor a loop in currency exchanges.

In turn, the behavior of charged objects gets modified by the non-trivial structure of the internal space.[20]

[20] In our financial toy model, the behavior of traders carrying copper gets modified as soon as arbitrage opportunities show up.

The picture that emerges this way is the following. Particles carrying electric charge modify the structure of the internal space which we call charge space. In turn, all electrically charged particles feel the structure of this internal space. Maybe it helps to imagine that our internal space gets curved due to the presence of the electric charges and therefore the path of electrically charged particles gets modified similarly to how the path of a marble is different when it rolls on a curved surface as opposed to when it rolls on a flat surface. And this is really how electrically charged particles influence each other. In other words, this is how electromagnetic interactions work. Electrically charged particles do not bump into each other. Instead, they respond to each other more indirectly by modifying the structure of charge space.[21]

[21] Again, don't worry if this is not perfectly clear at this point. We will talk about all the ideas mentioned here multiple times below.

While this story may still seem somewhat abstract and confusing, it is worth thinking about it in detail because all fundamental interactions work like this. The story is always the same, only the details are different.

Now, how can we implement the idea outlined above in quantum mechanics?

4.3 Interactions in Quantum Mechanics

Luckily, everything works completely analogously to what we already discussed in Section 2.2.1 for the financial toy model.

The main idea is once more that if there is a non-zero electric field strength $F_{\mu\nu} \neq 0$, we must be careful whenever we calculate derivatives. This is necessary because when we calculate the derivative of a function, we compare the values of the function at two different locations. And as soon as there is a non-zero electromagnetic field strength $F_{\mu\nu} \neq 0$, we know immediately that there will be a non-trivial electromagnetic potential A_μ somewhere. But when we want to compare the value of a function at two locations, we need to take this potential into account. This is analogous to how, in the financial toy model, the exchange rates are essential in determining the flow of copper (copper current) between neighboring countries.

There is only one small difference which has to do with the imaginary unit i in the exponent of our gauge transformations in quantum mechanics, $e^{i\epsilon(x)}$. To take the complex structure of the internal space of quantum mechanics into account, we have to add an imaginary unit to our covariant derivative formula.[22]

To cut a long story short: to take the effects of a non-zero electromagnetic field strength into account, again we need to replace our ordinary derivatives with **covariant derivatives**

$$\partial_\mu \to D_\mu \equiv \partial_\mu - iA_\mu \,. \tag{4.21}$$

In other words, this is what the derivative of a function looks like in the presence of a non-trivial electromagnetic

[22] One way to understand this is by noting that products of the form $\psi_i^\star(x)\partial_x\Psi(x)$ are not invariant under local transformations

$$\Psi(x) \to e^{i\epsilon(x)}\Psi(x)$$

$$\psi_i^\star(x) \to e^{-i\epsilon(x)}\psi_i^\star(x)$$

as we can check explicitly:

$$\begin{aligned}
&\psi_i^\star(x)\partial_x\Psi(x) \to \\
&\psi_i^\star(x)e^{-i\epsilon(x)}\partial_x\Big(e^{i\epsilon(x)}\Psi(x)\Big) \\
&= \psi_i^\star(x)\partial_x\Psi(x) \\
&\quad + \psi_i^\star(x)\Big(i\partial_x\epsilon(x)\Big)\Psi(x) \\
&\neq \psi_i^\star(x)\partial_x\Psi(x) \,.
\end{aligned}$$

However we can get something invariant by replacing the derivative ∂_x with the appropriate covariant derivative

$$D_x = \partial_x - iA_x,$$

where the bookkeeper A_x adjusts automatically

$$A_x \to A_x + \partial_x\epsilon(x)$$

whenever we perform a transformation of the wave function (see Eq. 2.21). Let's check this explicitly:

$$\begin{aligned}
&\psi_i^\star(x)(\partial_x - iA_x)\Psi(x) \to \\
&= \psi_i^\star(x)e^{-i\epsilon(x)}\Big(\partial_x - iA_x \\
&\quad - i\partial_x\epsilon(x)\Big)e^{i\epsilon(x)}\Psi(x) \\
&= \psi_i^\star(x)\partial_x\Psi(x) \\
&\quad + \psi_i^\star(x)\Big(i\partial_x\epsilon(x)\Big)\Psi(x) \\
&\quad - i\psi_i^\star(x)A_x\Psi(x) \\
&\quad - i\psi_i^\star(x)\Big(\partial_x\epsilon(x)\Big)\Psi(x) \\
&= \psi_i^\star(x)(\partial_x - iA_x)\Psi(x) \,.
\end{aligned}$$

potential A_μ.

Therefore, the Schrödinger equation (Eq. 4.2)

$$\partial_0 \psi(t, \vec{x}) = iD\partial_i^2 \psi(t, \vec{x}) \tag{4.22}$$

becomes

$$\Big(\partial_0 - iA_0\Big)\psi(t, \vec{x}) = iD\Big(\partial_i - iA_i\Big)^2 \psi(t, \vec{x}) \,. \tag{4.23}$$

This is what the Schrödinger equation looks like in the presence of a non-zero electromagnetic potential. It's really that simple and completely analogous to what we discussed for the financial toy model.

A small detail that we need to talk about before we move on is that different particles experience the influence of an electromagnetic field differently. To take this into account, we introduce a new constant q in front of the electromagnetic potential

$$\Big(\partial_0 - iqA_0\Big)\psi(t, \vec{x}) = iD\Big(\partial_i - iqA_i\Big)^2 \psi(t, \vec{x}) \,. \tag{4.24}$$

This new constant q is known as the **electric charge** of the particle in question. The constant q is a measure of how strongly the particle reacts to the presence of the electromagnetic potential A_μ. [23]

[23] In the financial toy model, q encodes how strongly traders react to the prices of copper and arbitrage opportunities.

Now, before we move on and discuss the remaining two internal spaces (isospin space and color space) and spacetime itself, it may be useful to take a step back and read Appendix D. In this appendix, everything we discussed so far is rephrased a bit more systematically using the proper mathematical notions. But, of course, you can also move on to the next chapter and read Appendix D afterwards.

In either case, we are ready to move on to the remaining two internal spaces.

4.4 Isospin Space and Color Space

Unfortunately, to properly describe isospin space, color space and the corresponding weak and strong interactions, we need quantum field theory. This is because we need to be able to describe how different types of particles can pop in and out of existence. And to describe systems in which the number of particles is not conserved, we need the formalism of quantum field theory.[24]

[24] To accurately describe electromagnetic interactions at high energies, we also need quantum field theory because then the particle number is no longer conserved. But for electromagnetic interactions, we can develop a theory which is valid at lower energies (this is what we did above) while for strong and weak interactions this is not possible because they are short ranged and therefore their effect is only visible at high energies.

Nevertheless, at this point we are capable of getting at least a glimpse of what is going on.

In the previous sections, we discussed how we can understand electromagnetic interactions by making use of the internal space we call charge space. The main idea there was that particles which carry electric charge curve this charge space and in turn, are influenced by the curvature of the charge space. This is how electromagnetic interactions take place.

In this section, we will do exactly the same thing again but this time we will understand how we can describe weak interactions using isospin space and how we can describe strong interactions using color space. In fact, it's really only the space which is different. The main ideas stay the same.

This means that particles carrying isospin curve isospin

space and in turn, are influenced by the curvature of isospin space. This is how weak interactions between elementary particles take place. Similarly, particles carrying color curve color space and feel the structure of color space. This is how strong interactions take place.

Therefore, our only task is to understand isospin space and color space. We just need to recall what we discussed when we talked about charge space and then go one step beyond that.

But before we dive in, here's one short comment. A question you may have is: why exactly these spaces? Unfortunately, the answer is: nobody knows yet. This is an open question in physics. We only know that these spaces are the right ones because we can observe their implications in experiments. But we don't know why only these specific structures show up.[25]

[25] A popular idea among theoretical physicists is to add new internal spaces or to unify the known spaces into one higher-dimensional internal space. These ideas are known as extended gauge models and Grand Unified Theories.

Now, what is isospin space and color space?

In the previous sections, we talked about internal spaces first and only afterwards about how they are related to interactions. But isospin space and color space are a bit more abstract and it is convenient to discuss the corresponding interactions first.

A first crucial fact we need is that in quantum theories there is not only a fundamental randomness in the movement of particles but also in their interactions. In a quantum theory, we can only calculate the probability that a certain interaction takes place. This is analogous to how we were previously only able to calculate the probability to find a certain amount of copper or electric charge at a particular location. But the randomness in the movement of particles and the randomness in interactions are, of course, not completely independent. How strongly two elementary particles interact depends crucially, for example, on the distance between them. And, if the movement of the particles is random, we can only calculate the probability to find them at a particular distance from each other. Therefore, we can't be certain about the strength of the interaction.[26]

[26] In the illustration, an electron performs a random walk before it interacts with an electron-neutrino, ν_e. The dashed line indicates the interaction and the shaded area illustrates the region in which an interaction becomes highly probable.

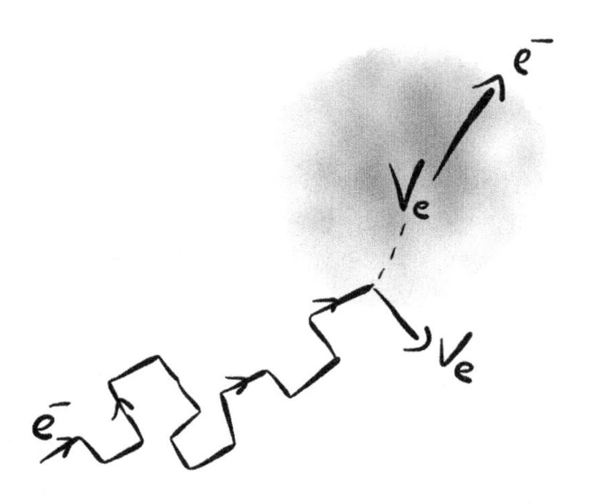

A second crucial fact which led to the discovery of isospin space and color space is that weak and strong interactions treat certain particles as if they were merely different states of the same particle. For example, weak interactions are blind to the electric charge of particles and cannot distinguish between an electron and an electron-neutrino; thus, in the internal isospin space, these two particles are perceived as different states of a single particle.[27] Similarly, strong interactions treat red up-quarks, blue up-quarks and green up-quarks as different states of the same object.[28] For the second example, this seems a bit more obvious thanks to the name choices, but the logic is really exactly the same.

[27] We will discuss what this really means in a moment.

[28] The color names red, blue, green have nothing to do with the colors we see in everyday life and are only used as convenient labels. Richard Feynman famously called his colleagues "idiot physicists" for choosing this confusing name, "which has nothing to do with color in the normal sense.".

If we combine these two observations, we can understand what color space and isospin space are.

The meaning of isospin space is that it encodes how much of a given state is an electron and how much of it is an electron-neutrino.

In other words, if there is an electron/neutrino state in our system, the corresponding axes of isospin space encode the "electron-ness" and "neutrino-ness" of the state.

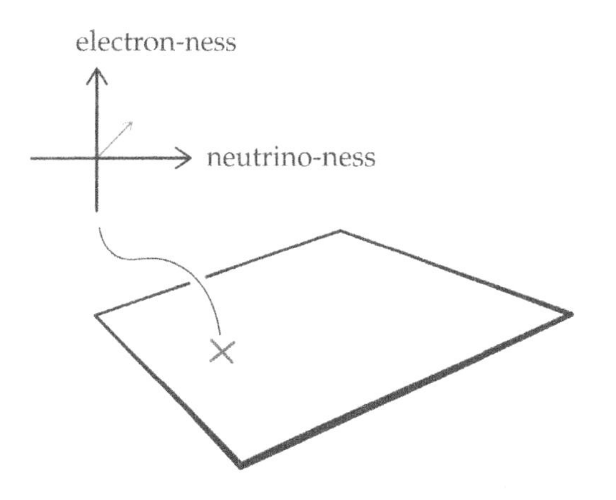

Similarly, if there is an up-quark/down-quark state in our system, the axes of isospin space describe the "up-ness" and "down-ness" of the states.[29]

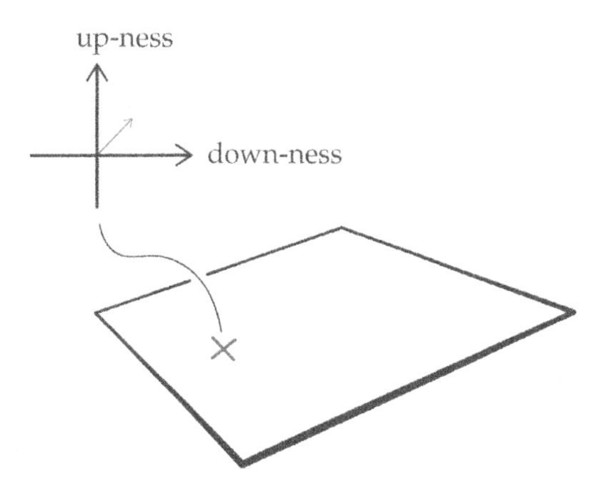

[29] Maybe it would help to call the electron-neutrino an up-electron and the ordinary electron a down-electron. Similarly, we could call a muon-neutrino an up-muon and the regular muon a down-muon, and a strange-quark an up-charm quark and the regular charm quark a down-charm quark etc. This way, it would become clear that we are dealing with different states of the same thing. We could then say that the axes of isospin space in general describe the relative "up-ness" and "down-ness" of the state at hand. The color names described below are a bit more helpful in this regard. However, we don't use these kinds of names for the neutrinos etc. because historically the isospin structure was discovered much later than most of the fermions.

Above each location, we have a copy of isospin space, analogous to how we have a copy of price space in the financial toy model. The total isospin space consists of all these local internal isospin spaces glued together. A specific state of our system corresponds to exactly one point in each of these spaces. A specific point in isospin space corresponds to a specific probability that we find an electron or an electron-neutrino at a given location.

Analogously, we have above each location a copy of color space and the total color space consists of all these local internal color spaces taken together. The axes of this space encode the redness, blueness and greenness of the state in question.

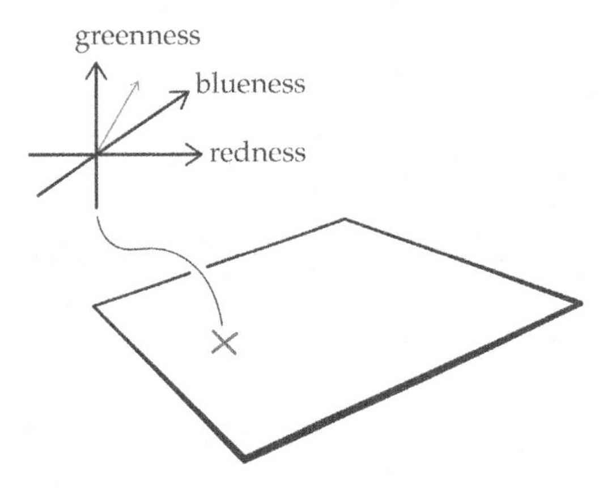

A specific point in color space corresponds to specific probabilities to find, say, a given up-quark as a red up-quark, blue up-quark or green up-quark.

This is certainly somewhat confusing and it takes some time to get used to this kind of thinking. But we will see in a moment that the rest of the story is completely analogous to what we discussed previously.

4.4.1 Symmetries of Isospin and Color Space

First of all, as soon as we are dealing with a space of any sort, we can ask: what are its symmetries? The most natural assumption is always that each space possesses symmetries. We saw this previously for the price space and for charge space. And again the same is true for isospin space and color space. The symmetries of charge space are simply rotations in the complex plane. Mathematically, we describe these rotations using the group $U(1)$.[30]

Now, the symmetries of isospin space and color space are, in some sense, simply higher-dimensional analogs of these rotations. The symmetry group of isospin space is known as $SU(2)$ and the symmetry group of color space is $SU(3)$.[31]

[30] This is discussed in a bit more detail in Appendix D.

[31] $SU(2)$ consists of all unitary (2×2) matrices with unit determinant. The S denotes "special" which means determinant one. The U denotes "unitary", which means that the group elements fulfill the condition $g^\star g = 1$. Analogously, $SU(3)$ consists of all unitary (3×3) matrices with unit determinant. In this sense, these symmetries are the next simplest thing we can try after $U(1)$. (Note that the group $SU(1)$ also exists but consists only of the trivial element 1 which does nothing at all.) Discussing group theory and these groups in detail would lead us too far astray, but, if you're intersted, you can find a detailed discussion in

Jakob Schwichtenberg. *Physics from Symmetry*. Springer, Cham, Switzerland, 2018b. ISBN 978-3319666303

4.4.2 Isospin and Color Connections

Again, the key idea is that we allow arbitrary local conventions. In the finance toy model, we allowed countries to use local currencies. Here, we allow each location to choose freely which axis is called the electron axis and which is called the neutrino axis. These are just labels and local choices do not matter as long as we take them into account consistently. This is again where bookkeepers become important. In the context of weak interactions, it is conventional to use the symbol W_μ for the bookkeepers.

While the local choice between electron and neutrino axes doesn't make any difference, we need to be careful when we compare states at two different locations. The relative "electron-ness" and "neutrino-ness" between the states at two locations is important and cannot depend on local conventions.[32]

[32] This is analogous to how a price difference in two different countries is something physical. (At least, if we convert the prices to a common currency before we compare them.)

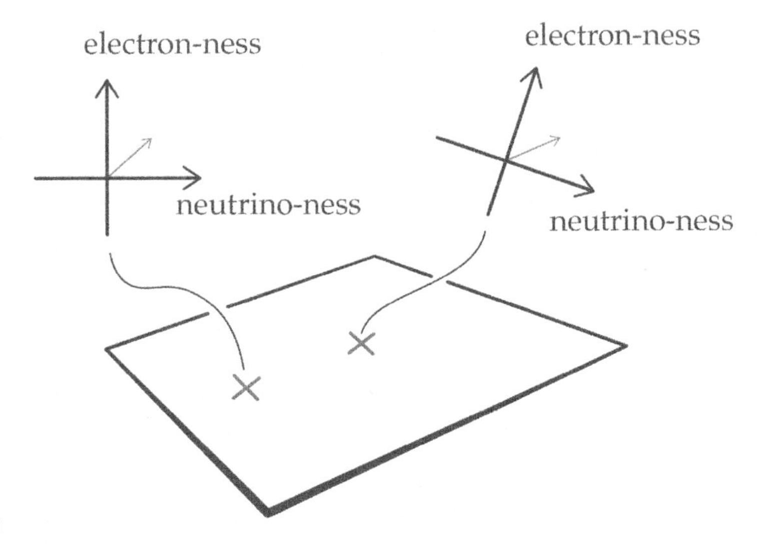

The bookkeepers W_μ keep track of the local conventions and always adjust so that the relative "electron-ness" and "neutrino-ness" stay unaffected by local conventions.[33]

[33] There is an important point we are glossing over here. Since our internal spaces are now higher-dimensional, we need multiple types of bookkeepers. For isospin space, we need three bookkeepers $W^1_\mu, W^2_\mu, W^3_\mu$. In mathematical terms, this follows because the adjoint representation of $SU(2)$ is three-dimensional.

Formulated a bit more technically, the bookkeepers W_μ are (again) the connections which allow us to consistently transport a given state to a different location.[34] This transportation is necessary whenever we want to compare a state at one location with another state at another location.[35]

[34] The mathematical concept of *connection* is discussed in more detail in Appendix D.4.

[35] For example, whenever we calculate the derivative.

We can't compare them directly because the description of states can depend on local conventions. But, by transporting one state to the location of a second state, we make sure that we compare them using the same local conventions.

Analogously, the redness, blueness and greenness of a given state can be redefined arbitrarily at each location.

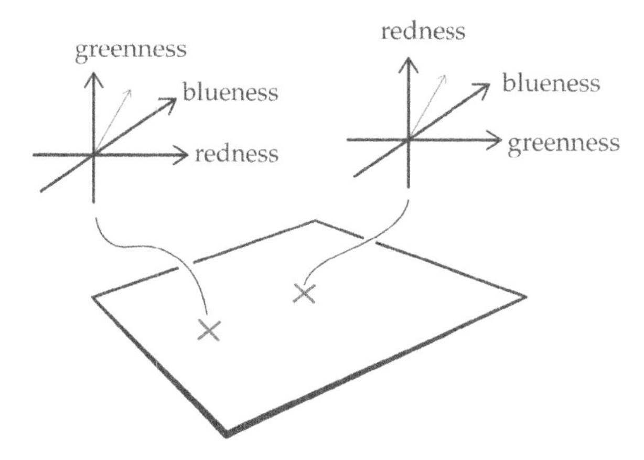

But if we want to compare states at two different locations, the relative redness, blueness and greenness become important.

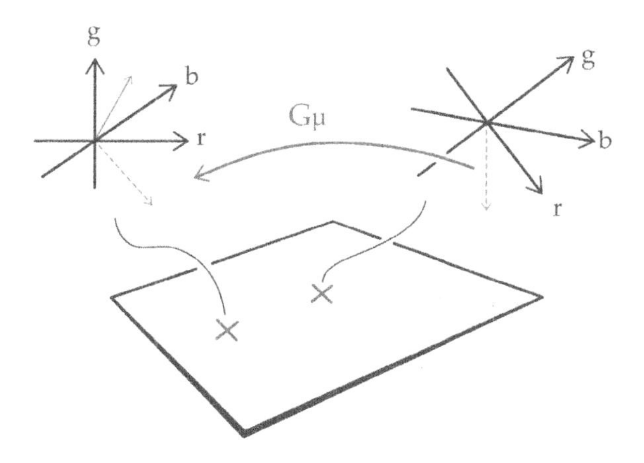

Therefore, we once more need bookkeepers G_μ to keep track of these local choices.[36]

The final puzzle piece we need to talk about is what all this really has to do with weak and strong interactions.

[36] Color space is a lot more complicated than isospin space and we need eight bookkeepers to keep track of all local conventions. In mathematical terms this follows because the adjoint representation of $SU(3)$ is eight-dimensional.

4.4.3 Weak and Strong Interactions

Now, the main idea is once more completely analogous to what we discussed in the financial toy model and for electromagnetic interactions. The freedom to choose local conventions is a nice feature but not really essential. Nevertheless, it is possible that our bookkeepers become indispensable. In the financial toy model this is the case when there are currency arbitrage opportunities ($F_{\mu\nu} \neq 0$). The existence of such an opportunity is a real feature of the model which influences the dynamics and implies directly that there has to be a non-trivial exchange rate A_μ somewhere. Analogously, if there is a non-zero electromagnetic field strength in the model $F_{\mu\nu} \neq 0$, we know immediately that the electromagnetic potential A_μ has to be non-vanishing somewhere.

Completely analogously, our isospin bookkeepers W_μ and color bookkeepers G_μ become indispensable when there is a non-zero weak field strength and a non-zero strong field strength somewhere.[37] Don't let the somewhat unfortunate names "weak" and "strong" confuse you. We are really dealing with new types of interactions and not simply weak electromagnetic or strong electromagnetic interactions. For example, the weak force is responsible for the decay of nuclei while the strong force holds protons and neutrons together.[38] These are phenomena which cannot be explained by electromagnetic interactions.

There are particles which can modify the structure of the total isospin and color space such that non-zero field strengths show up. We say that a particle which modifies the structure of isospin space carries isospin. A particle which modifies color space carries color. Again, this is analogous to how only particles carrying electric charge modify the struc-

[37] Take note that only for electromagnetic interactions does the field strength tensor have the simple form (Eq. 2.22)

$$F_{\mu\nu} \equiv \frac{\partial A_\nu}{\partial x^\mu} - \frac{\partial A_\mu}{\partial x^\nu}.$$

The field-strength tensor for weak and strong interactions contains an additional term:

$$W_{\mu\nu} \equiv \frac{\partial A_\nu}{\partial x^\mu} - \frac{\partial A_\mu}{\partial x^\nu} + gT_a f^a_{bc} A^b_\mu A^c_\nu,$$

where g is a constant which characterizes the strength of the corresponding interaction, f^a_{bc} are the so-called structure constants and T_a is a generator of the corresponding symmetry group. This term was only absent for the electromagnetic field-strength tensor $F_{\mu\nu}$ because the structure of charge space is extremely simple. Mathematically, we get an extra term like this whenever the symmetry group is non-abelian. Physically, this extra term describes self-interactions of the corresponding gauge bosons. We will see something very similar when we discuss the tensor describing gravity.

[38] Protons and neutrons consist of quarks.

ture of charge space. And only particles which carry electric charge, isospin charge or color charge are influenced by the structure of the corresponding space.

This is how particles interact with each other.[39]

[39] Hopefully this diagram starts to make a bit more sense.

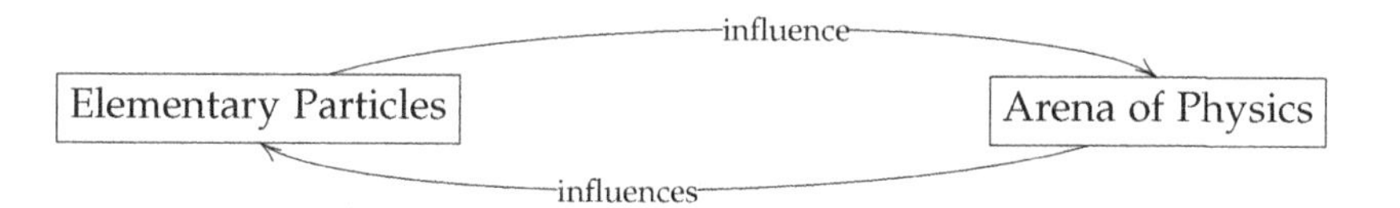

But this geometrical picture is not only important because it lets us understand all these fundamental interactions from a common perspective. We can also use it to understand an important feature of weak and strong interactions.

To understand this feature, we need to ask: what does it really mean that bookkeepers are indispensable here?

Let's assume we are 100% certain that we are dealing with an electron at a particular location. Moreover, let's assume that we have fixed all local conventions and we try to keep them as simple as possible.[40] The main point is that if there is a non-zero weak field strength somewhere in the system, we necessarily have a non-trivial bookkeeper W_μ. That is, even if we try to choose our local isospin spaces extremely cleverly such that our pure electron state will correspond to a pure electron state everywhere (if we move it around in time and space), we will fail because otherwise the bookkeeper W_μ would be zero everywhere. We know that W_μ has to be non-zero somewhere since there is a non-vanishing field strength. This implies that even if we start with a pure electron state and try to make sure that it corresponds to a pure electron state everywhere in time and space, we will get a state with a non-vanishing electron-

[40] In the financial toy model this would mean that we try to use the same (global) currency everywhere. However, this turns out to be impossible if there is an arbitrage opportunity since this means that there must be a non-trivial exchange rate somewhere.

neutrino component somewhere.

In physical terms, this means that a pure electron state can be transformed into an electron-neutrino state through weak interactions. Formulated more precisely, if there is a non-zero weak field strength, there is a non-zero probability to find that our electron has turned into an electron-neutrino at a later point in time.

Do you wonder about the conservation of electric charge? After all, an electron carries electric charge while an electron-neutrino does not. Rest assured: in all particle interactions observed so far, electric charge was found to be conserved. Therefore, if our electron is transformed into a neutrino, the electron's electric charge must go somewhere.

To understand how this works, we need to recall that the non-zero field strength which affects the electron in question is caused by a second particle. But, as usual in physics, this second particle is, in turn, influenced by the electron.[41] This second particle is exactly where the electron's electric charge ends up.

[41] action = reaction

The picture which emerges this way is the following. We start with two particles: an electron at some location A and, say, an electron-neutrino at a different location B. Each of these two particles modifies the structure of isospin space and is influenced by it.

This is how they "talk" to each other. What we discussed above means that there is a non-zero probability that after some time we will find an electron-neutrino at location A and an electron at location B. [42]

[42] Take note that this is only possible if our connections are able to carry electric charge. And this is indeed what we observe in nature. Two of the three particles associated with weak interactions do indeed carry electric charge. This also indicates that there is a close connection between weak and electromagnetic interactions. In fact, in modern physics we usually talk about electroweak interactions. To properly understand this requires an understanding of spontaneous symmetry breaking which is somewhat beyond the scope of this book.

space
time
e^- ν_e W^- ν_e e^-

This is a real physical process mediated by weak interactions which we can observe in nature.

Completely analogously, an up-quark can turn into a down-quark and vice versa.

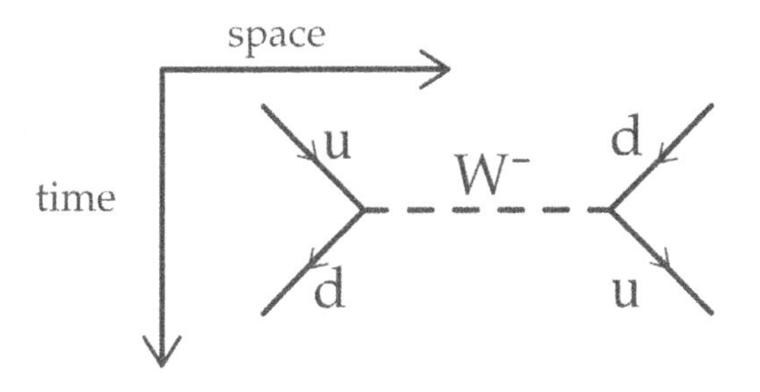

And exactly the same picture emerges if we consider strong interactions. The only difference is that it's the "color" of the quarks which gets modified.[43] Specifically, this means, for example, that a red up-quark gets transformed into a blue up-quark or a green up-quark. But again, we must make sure that everything important is conserved. This means that color charge must go somewhere and a different quark at some other location must therefore provide the required color and absorb the color which our first quark "lost".

[43] The logic is really the same as before. The main reason why we treat different states related to weak interactions as truly different particles is that the corresponding symmetry ($SU(2)$) is broken in nature. As a consequence, the two states inside each doublet (the electron-neutrino and the electron or the up-quark and the down-quark) have different masses. This is why we give these states completely different names. In contrast, the symmetry associated with strong interactions ($SU(3)$) is intact, which is why we treat the different color states as different versions of the same particle. (A green quark has the same mass as a blue quark and a red quark.)

Take note that the various interactions are not mutually exclusive. Instead, they happen simultaneously all the time. A particle which carries electric charge, isospin charge and color charge (e.g. a red up-quark) is influenced by the structure of color space, isospin space and charge space simultaneously! Therefore, if we want to calculate the probability of finding a given particle in a particular final state, we need to take lots of things into account. As mentioned before this is possible using quantum field theory. One of the main goals in quantum field theory is indeed to calculate the probabilities of finding a given set of particles in a different state after some time.[44]

[44] Unfortunately, a proper discussion of quantum field theory is far beyond the scope of this book. But if you want to learn more about it, you might enjoy my book

Jakob Schwichtenberg. *Physics from Symmetry*. Springer, Cham, Switzerland, 2018b. ISBN 978-3319666303

There are two final comments to make before we move on to the final part of the arena of physics and the interaction associated with it.[45] Firstly, maybe you wonder, since the effect of weak and strong interactions is to turn particles into different particles, what exactly do electromagnetic interactions do?

[45] Spacetime and gravity!

The answer is: electromagnetic interactions change the momentum of particles.[46] If we start with two electrons, each with some given momentum, we end up again with two electrons but they possibly carry different momenta.

[46] The connection between charge space, which encodes the freedom to rotate the phase of wave functions, and momentum can be understood as follows. The momentum of a particle described by a wave function $\psi = Ae^{i\varphi}$ is directly proportional to the derivative of the phase φ of the wave function. If there is a non-zero electromagnetic field, the momentum of the particle becomes proportional to $\partial_\mu \varphi - qA_\mu$ where q denotes the electric charge of the particle.

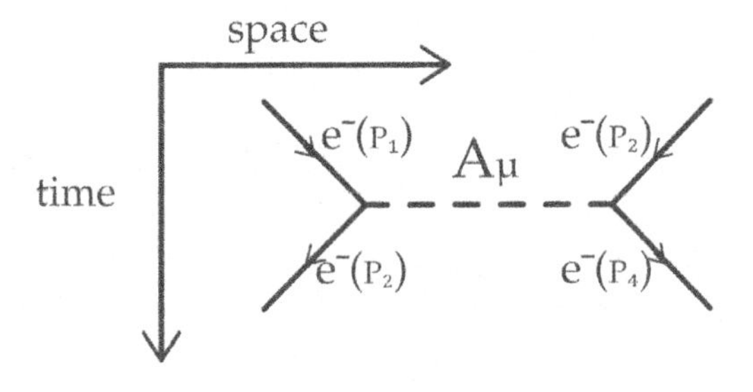

The momentum one particle loses is transported via the

electromagnetic connection A_μ to the location of the second particle.[47]

Similarly, when an electron scatters with a muon, we again get as a final state an electron plus a muon but with different momenta.

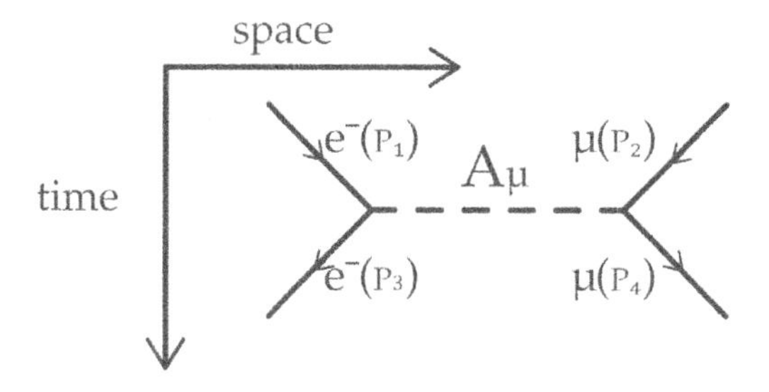

And, of course, electrons and muons can also interact electromagnetically with quarks since all these particles carry electric charge:

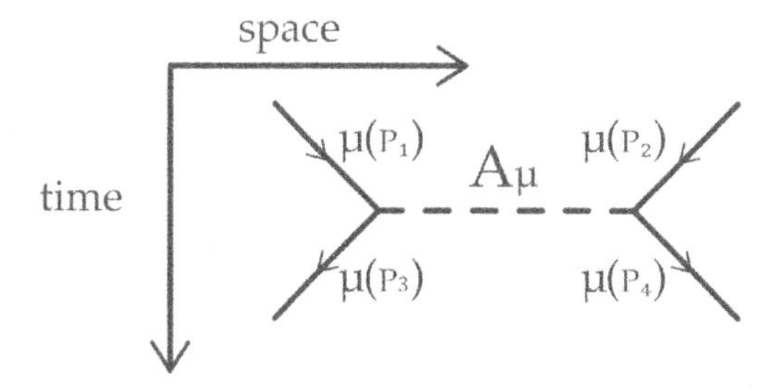

Secondly, maybe you wonder why we only observe electromagnetic interactions in everyday life but never weak and strong interactions? The reason for this is that strong and weak interactions are extremely short-ranged while electromagnetic interactions are long-ranged. This means that a

[47] Also take note that the momenta carried by the final particles in weak and strong interactions are, in general, different from the momenta carried by the initial particles. Yet, the total momentum is always conserved.

particle which carries isospin only influences isospin space in a small region of spacetime around it. Similarly a particle which carries color charge only influences color space in a tiny neighboring region of spacetime.[48]

[48] This feature of weak interactions can be understood because the corresponding gauge bosons are massive. (In the pictures shown above you can imagine that the line between the electron and the neutrino represents an additional particle W_μ which mediates the interaction between the two.) This is a result of the spontaneous symmetry breaking of the corresponding $SU(2)$ symmetry through the Higgs mechanism. As a result, the massive gauge bosons cannot travel very far before they decay into lighter particles and that's why weak interactions are short-ranged. In contrast, the gauge bosons mediating strong interactions are massless. Nevertheless, strong interactions are short-ranged because colored objects always lump together to color-neutral objects. This mechanism is known as confinement but there is still no completely satisfactory explanation for why and how this happens.

In contrast, every particle which carries electric charge influences charge space in a huge region around it. This is except, of course, when there is another particle nearby with exactly the opposite charge. Then, their effects on charge space cancel each other. This happens, for example, because the charge of an electron is $-e$ and the charge of a proton is $+e$. Thus, we get an electrically neutral composite object whose combined effects on charge space cancel.

5

Special Relativity and General Relativity

First of all, a short disclaimer. As before, we will only focus on very special aspects of special relativity and general relativity. There is, of course, a lot more one could and should say about them. But discussing the nuances of special and general relativity requires at least a few hundred pages and is thus not something we can realistically do here. We will skip many mathematical details and lots of important concepts are not mentioned at all. Nevertheless, the main features of both theories can be understood beautifully using the same ideas discussed already in the previous chapters. And this is the only thing we will focus on.

So far, we have talked about internal spaces and how we can use them to understand how particles interact with each other. The main idea was that particles which carry charges modify the structure of the corresponding internal spaces. At the same time, each such particle is influenced by the structure of the internal spaces.[1] This is how particles interact even though they do not directly touch each other.

Now, there is exactly one additional known space (spacetime) and exactly one additional known fundamental interaction which we haven't talked about so far. Gravity.[2]

At this point you will probably not be surprised if I tell you that there is a direct connection between the structure of spacetime and gravity, completely analogous to how there is a direct connection between the structure of charge space and electromagnetic interactions. But historically, understanding this connection was a huge intellectual step forward and is rightfully regarded as one of the most beautiful insights in physics. It took a genius like Einstein more than a decade to figure it out. Nowadays, with the power of hindsight, we can understand it almost effortlessly in just a few minutes.

In fact, not only is there a close connection between gravity and spacetime, but we can understand this interplay completely analogously to how we have come to understand the connection between the other fundamental interactions and their corresponding (internal) spaces.

The main idea is once more that objects carrying the corresponding charge, here energy-momentum, modify the structure of spacetime. In turn, all objects carrying energy-momentum are influenced by the structure of spacetime. This is how gravitational interactions take place!

Moreover, analogous to how a non-zero electromagnetic field strength corresponds to a non-zero curvature of charge space, a non-zero gravitational field strength corresponds to a non-zero curvature of spacetime.

[1] To be influenced by the structure of a given internal space, a particle must carry the charge that is associated with that internal space. For example, only electrically charged particles are influenced by the structure of charge space. This is discussed in a bit more detail in Appendix D.

[2] There could, of course, be additional internal spaces and fundamental interactions. But so far, only gravity, electromagnetic, weak and strong interactions have been observed in experiments.

In other words, while a few details are different, the general story-line is exactly the same.[3]

To understand all this a bit better, we start by discussing the local structure of the space in question.[4] Afterwards, we again need something ("bookkeepers") to glue these individual local spaces together. Then we can talk about situations in which our individual local spaces are glued together non-trivially. As before, such situations are characterized by a non-zero curvature. But now we will talk about the curvature of spacetime and not that of some internal space. Finally, we will talk about the Einstein equation which describes how a non-zero curvature of spacetime arises, analogous to how Maxwell's equations describe how a non-zero curvature of charge space arises whenever electric charges are present.

[3] Take note that even though general relativity is more than a century old, experts still argue about basic concepts; like which geometric structure correctly describes our universe. But like the discussion about the correct interpretation of quantum mechanics, this is not something we will discuss here.

[4] The local structure in the financial toy model was simply a line, while for quantum mechanics our local internal space is a circle.

With this plan in mind, let's dive in.

5.1 Special Relativity

What do we know about the structure of space and time?

Maybe this seems like a strange question, but there are many important things we can learn by answering it.

First of all, we usually think of the space we live in as a relatively boring arena. To describe the location of an object, we introduce a coordinate system and determine three numbers (x, y, z). These numbers indicate the location of the object with respect to three coordinates axes.

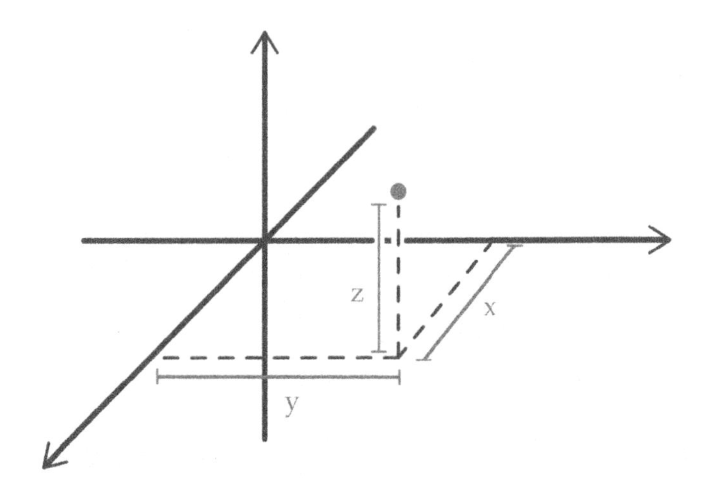

Since, in general, we need three such numbers we say that space is three-dimensional. And as you can easily check with a ruler, the distance Δs between any two objects in this three-dimensional space can be calculated using the Pythagorean theorem[5]

$$\Delta s^2 = \Delta x^2 + \Delta y^2 + \Delta z^2 \,. \tag{5.1}$$

[5] From another perspective, we can say that we calculate the length of the vector

$$\vec{v} = \begin{pmatrix} \Delta x \\ \Delta y \\ \Delta z \end{pmatrix}$$

pointing from one point to another:

$$|\vec{v}|^2 = \vec{v} \cdot \vec{v} = \Delta x^2 + \Delta y^2 + \Delta z^2 \,.$$

Since we can calculate distances like this, we say that the space we live in is Euclidean.[6]

[6] Maybe you wonder why we care about such a trivial fact and even introduce a special name? We will see in a moment that the above relation is not always true, i.e. there can be non-Euclidean spaces.

Now, in physics we are not only interested in knowing *where* something happens but also *when*. After all, one of our main goals in physics is to describe how objects move around. This requires that we take time into account.

Therefore, to properly describe an object in physics, we actually need four numbers (t, x, y, z). The time coordinate t tells us the time at which the object is at the location (x, y, z). This means we add a fourth axis to our coordinate system. In other words, we now not only use a spatial coordinate system, but also a **spacetime** coordinate system. While (x, y, z) describes the location of an object in space, we say that (t, x, y, z) describes the location of an **event** in spacetime. And since we need four numbers to describe the location of an event, we say that spacetime is four-dimensional.

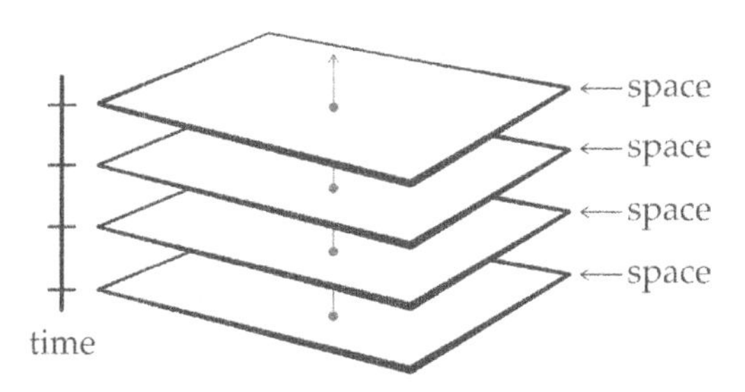

Hopefully you are not too bored or confused by these lines of thought because things are about to get really interesting.

We defined above the spatial distance between two objects (Eq. 5.1). Now that we've added time to our coordinate system, we can ask: what's the distance between two events in spacetime? Naively, we might write down

$$\Delta s^2 = \Delta t^2 + \Delta x^2 + \Delta y^2 + \Delta z^2 \tag{5.2}$$

but this doesn't make any sense. The differences in the spatial components $(\Delta x, \Delta y, \Delta z)$ are measured in meters, while Δt is measured in seconds. This means we are comparing apples with oranges in Eq. 5.2. However, we can fix this problem by introducing a new constant c which has units meters per second:

$$\Delta s^2 = c^2 \Delta t^2 + \Delta x^2 + \Delta y^2 + \Delta z^2 \,. \tag{5.3}$$

Now, the first term in the sum on the right-hand side has units $\left(\frac{\text{meters}}{\text{second}}\right)^2 \text{seconds}^2 = \text{meters}^2$ as it should be.[7]

[7] We will talk about the meaning of this constant in a moment.

One of the big discoveries in physics was that this formula is *not* the relevant one.

Instead, the relevant expression for the (squared) "distance" between events is[8]

[8] Historically, the Michelson-Morley experiment showed the constancy of the speed of light, independent of the reference frame in which it is measured. This fact, which lies at the heart of special relativity, was formalized by Einstein and Minkowski. But before we address this, we need to understand the physical implications of this relation.

$$\Delta s^2 = c^2 \Delta t^2 - \Delta x^2 - \Delta y^2 - \Delta z^2 . \tag{5.4}$$

In words, this means there is a relative minus sign between the (squared) spatial distance $\Delta x^2 + \Delta y^2 + \Delta z^2$ and the (squared) temporal distance Δt^2.[9] While this formula may look extremely strange, from a mathematical perspective, it simply tells us that the local structure of spacetime is not what we would've naively expected. In some sense this is analogous to how we discovered that the local structure of charge space is not a simple line but a circle. Here, we discover that the local structure of spacetime is not Euclidean (Eq. 5.3) but Minkowskian (Eq. 5.4). The main feature of a Minkowski space compared to a Euclidean space is the way we define distances. Mathematically, we denote the four-dimensional Euclidean space as $\mathbb{R}^4$ and the four-dimensional Minkwoski space as $\mathbb{R}^{1,3}$.

[9] We will talk about the physical meaning of the minus sign in Eq. 5.4 in a moment.

But the minus sign not only encodes an important fact about the local structure of spacetime, it also has extremely important physical implications.

To understand this, let's consider two events called A and B which are spatially separated by 3 meters. Moreover, we assume that B is caused by A. For example, we can imagine that event A is the emission of a light pulse and event B is the detection at a location that is 3 meters away. Now, how soon after A can B happen? In other words, what's the minimum temporal distance Δt between the two events?

We can answer this question by looking at Eq. 5.4. The (squared) spatial distance between the two events $\Delta x^2 +$

$\Delta y^2 + \Delta z^2 = (3 \text{ meters})^2$ is fixed. Therefore

$$\Delta s^2 = c^2 \Delta t^2 - \Delta x^2 - \Delta y^2 - \Delta z^2$$
$$\quad \text{↷ } \Delta x^2 + \Delta y^2 + \Delta z^2 = (3 \text{ meters})^2$$
$$= c^2 \Delta t^2 - (3 \text{ meters})^2 . \tag{5.5}$$

Our goal is to find the minimum allowed value of Δt. A crucial observation is that Δt cannot be arbitrarily small and certainly cannot be zero; that is, if we forbid instantaneous action at a distance. Since the two events are causally related, the pulse of light necessarily needs some time (even if very short) to travel the spatial distance between the (separate) locations of emission and detection. Therefore, the minimum allowed value of Δt is precisely the value for which Δs^2 is zero.

The main point is that there is a *non-zero* minimum time period between two events which happen at two different locations. This implies that there is a **maximum velocity** at which a signal can travel from one point to another.

Using Eq. 5.4 we can learn something important about this maximum velocity. A signal which travels at the maximum velocity between the two events, needs the minimum time interval Δt to travel the distance $\Delta x^2 + \Delta y^2 + \Delta z^2$. Above, we argued that the minimum time interval Δt corresponds to $\Delta s = 0$. Putting this into Eq. 5.4 yields

$$\Delta s^2_{\text{min}} = 0 = c^2 \Delta t^2_{\text{min}} - \Delta x^2 - \Delta y^2 - \Delta z^2$$
$$\text{↷}$$
$$c^2 = \frac{\Delta x^2 + \Delta y^2 + \Delta z^2}{\Delta t^2_{\text{min}}} . \tag{5.6}$$

This is interesting because in general

$$v \equiv \frac{\sqrt{\Delta x^2 + \Delta y^2 + \Delta z^2}}{\Delta t} \tag{5.7}$$

is exactly the velocity at which a signal travels the distance $\sqrt{\Delta x^2 + \Delta y^2 + \Delta z^2}$.[10] Therefore, Eq. 5.6 tells us that a signal

[10] A velocity is always defined as a spatial distance divided by the time interval Δt needed to travel the distance.

which needs the minimum amount of time Δt_{min} travels at velocity c.

In other words, the constant c, which we introduced to get the same units for all terms in Eq. 5.4, encodes the maximum velocity at which anything can travel in spacetime. The constant c is usually called the **speed of light** because, well, light travels at this maximum velocity.[11] From experiments we know that while c is incredibly large, it's not infinite.[12] And this is historically how the correct local structure of spacetime (Eq. 5.4) was discovered.

The experimental fact that there is a maximum velocity which is valid for anyone and anything is the basis of Einstein's theory of special relativity.[13]

To summarize, the main idea described in this section is that there is an upper speed limit. This limit implies that the interplay between time and space works a little differently from what we would expect. Naively, we would expect that the distance between two objects is simply $\Delta s^2 = \Delta x^2 + \Delta y^2 + \ldots$ or $\Delta s^2 = (c\Delta t)^2 + \Delta x^2 + \Delta y^2 + \ldots$ if we are including a coordinate with different units like t.[14] However, if there is an upper speed limit, there is actually a relative minus sign between the temporal and spatial coordinates: $(\Delta s)^2 = (c\Delta t)^2 - (\Delta x)^2 - (\Delta y)^2 - \ldots$.[15] By including this relative minus sign, we are hard-coding the upper speed limit into the geometry of space and time.

We can visualize the structure of the resulting space as follows. At each moment in time we imagine that light (or anything else moving with the maximum velocity c) moves away from each particular location. The paths traced

[11] Take note that c is a general constant which often appears in contexts which have nothing to do with light. The name "speed of light" is only used for historic reasons. In general, c is an upper speed limit for everything in physics and all massless particles travel at speed c.

[12] The experimental value of c is

$$2.9979 \times 10^8 \, \frac{\text{meters}}{\text{second}}.$$

[13] Discussing special relativity any further here would lead us too far astray. But for the modest goal of this book, we already have everything we need to move forward. A few more mathematical details about the structure of Minkowski space are discussed in Appendix B.

[14] In mathematical terms, a space in which this is true is called an Euclidean space.

[15] The space in which we calculate distances this way is known as Minkowksi space.

this way define the boundaries of our arena. Every object starting at this moment in time and at this location has to remain within these boundaries. Nothing can move outside of this so-called **light cone** because this would require that it moves faster than c.

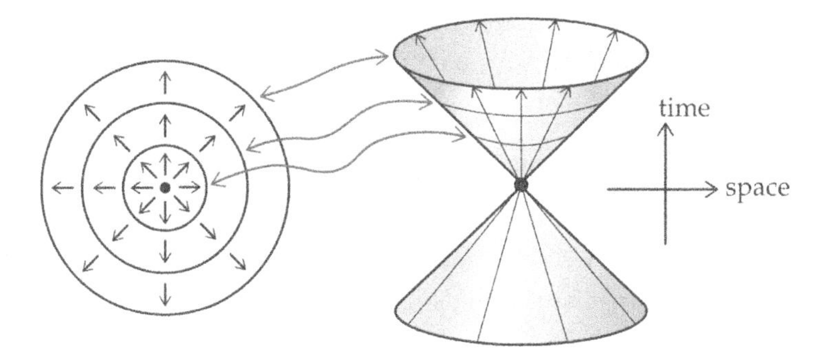

Since we get such a light cone for each possible location at each moment in time, we get the following picture:

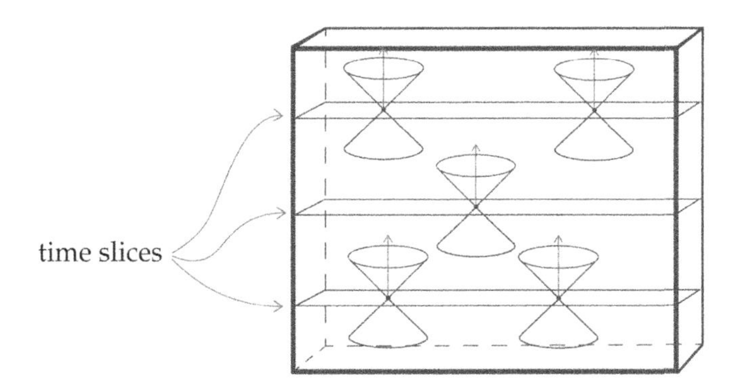

Before we move on and talk about the global structure of spacetime and general relativity, there is one additional thing worth mentioning. In the previous sections we noted that it can be instructive to analyze a given space using its symmetries. In particular, this is the case for isospin and color space. The structure of these spaces can be best understood by noting that the symmetries of local isospin and local color space are $SU(2)$ and $SU(3)$ respectively. Since Minkwoski space is also quite complicated, it also makes sense to study its symmetries to understand it better. One type of symmetry of spacetime is certainly rotations. But Minkowski space is four-dimensional and also includes a time-coordinate. This means there are also transformations which mix spatial and temporal components. These kinds of transformations are commonly called boosts. In physical terms, a boost is a transformation to a coordinate system moving at a constant velocity with respect to the initial reference frame. This is a symmetry because there is no way to tell whether you are sitting still or moving with a constant velocity. (For example, if you close your eyes and block out all noise in a stable train moving at constant velocity.)

Another symmetry of spacetime is translations. A spatial translation is a shift to another location $x \to x + \epsilon$ and temporal translation is a shift to a different point in time $t \to t + \epsilon$. The total symmetry group of Minkowski space therefore consists of rotations, boosts and translations. Mathematically, we call this symmetry group $P(1,3)$.[16]

Now, let's move on and discuss why Minkowski space is not enough to describe spacetime on a global scale.

[16] Mathematically, the Poincaré group $P(1,3)$ is the semidirect product of the Lorentz group $SO(1,3)$ and the translation group $\mathbb{R}^{1,3}$:

$$P(1,3) = \mathbb{R}^{1,3} \rtimes SO(1,3)\,.$$

The S denotes again "special" which means that all elements of $SO(1,3)$ have determinant one. The O denotes "orthogonal" and means that all $SO(1,3)$ elements fulfill the condition $O^T O = 1$. Take note that rotations in three-dimensional Euclidean space $\mathbb{R}^3$ are described mathematically by the group $SO(3)$. This interpretation comes about because transformations which fulfill the condition $O^T O = 1$ leave the length of vectors unchanged. Moreover, transformations with determinant one leave the orientation of the coordinate system unchanged. Therefore, taken together, these two conditions single out rotations from the set of all imaginable transformations. Now, we consider Minkowski space $\mathbb{R}^{1,3}$. "Rotations" in Minkowski space are denoted analogously $SO(1,3)$. Take note that translations in some space are mathematically equivalent to the space itself. For example, the group of translations in Minkowski space ($\mathbb{R}^{1,3}$) is $\mathbb{R}^{1,3}$. The Poincaré group is discussed in detail in my book *Physics from Symmetry*.

5.2 General Relativity

As mentioned at the beginning of this chapter, our goal is to repeat for spacetime what we discussed previously for the various internal spaces.

In the previous section, we discussed the local structure of spacetime. This step is analogous to how we discussed the local structure of the internal space in our financial toy model (a line) or in quantum mechanics (a circle). With this local structure at hand, we can follow exactly the same steps as in the previous chapters.

First of all, we can argue that it should be possible to use arbitrary local coordinate systems, analogously to how we argued that it should be possible to use arbitrary local currencies. For example, it shouldn't make any difference whether we use a rotated coordinate system at a later point in time or in a different region of spacetime. Maybe you think it's a somewhat silly idea to use different coordinate systems in different regions. But in some sense, this is exactly what we do whenever we use curvilinear coordinates. We need bookkeepers Γ as soon as we use curvilinear coordinates, e.g., spherical coordinates. When we use spherical coordinates the coordinate axes at different locations are orientated differently. This means that the components of a given vector change if we move it to a different location even though the vector itself doesn't change. Therefore, if we want to compare two vectors at two different locations, we can't compare their components directly. Instead, we need to move one vector to the location of the other vector. To do this consistently, we need a connection which keeps track of the local orientation of the coordinate axes throughout space.

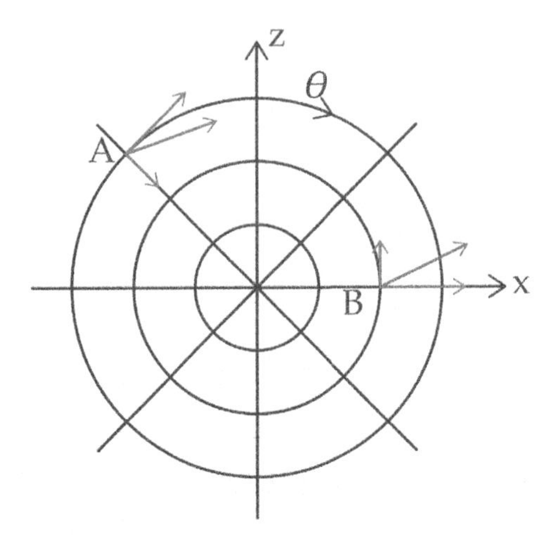

Once more, this is only possible if we introduce bookkeepers Γ which keep track of all local conventions. But again, this is not really essential *if* we can simply use the same choice of coordinate system everywhere; *unless*, of course, there is something which prevents this.

And indeed it turns out that this is not always possible, analogous to how in the financial toy model it's sometimes impossible to define a global currency. In the financial toy model, we discovered that whenever there is a currency arbitrage opportunity there can't be a global currency. A global currency would mean that we can eliminate all exchange rates A_μ. But this in turn would imply that there is no arbitrage opportunity $F_{\mu\nu}$.

Moreover, we argued that geometrically a currency arbitrage opportunity corresponds to a non-zero curvature of our internal money space.

One of the main ideas of general relativity is that exactly the same is true for spacetime.

If the curvature of spacetime is non-zero, it's impossible to use the same Minkowski coordinate system globally. In other words, we can't describe *global* spacetime using a single Minkowski space, but it is possible to neatly glue together multiples copies of *local* Minkowski spaces in a non-trivial manner.[17]

[17] The only exceptions where this type of construction doesn't work are regions where the curvature of spacetime gets infinite (ex: singularity at the center of a black hole, or the big bang singularity at the origin of the universe).

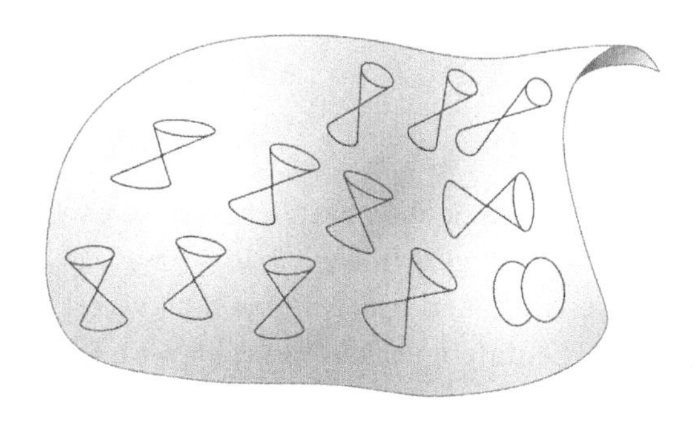

This is analogous to how the total internal money space consists of multiple lines which are glued together non-trivially.

Before attempting to understand what all this has to do with gravity, we need to talk about curvature. The main idea is that when we move an object along a closed path in a curved space, the final state of the object is not necessarily equivalent to its initial state.[18]

[18] This is discussed in detail in Appendix D.5.

The difference between the initial and the final state encodes how curved the space is.[19]

[19] Take note that this is exactly how we defined the gain factor $F_{\mu\nu}$ which indicates that there is a currency arbitrage opportunity.

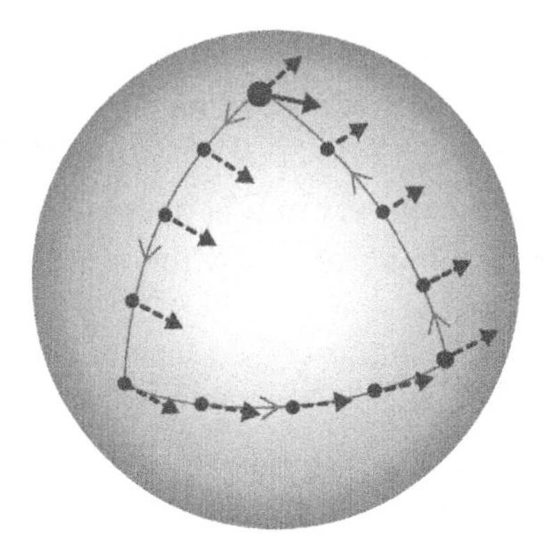

We use exactly the same idea to describe the curvature of spacetime. Specifically this means that we move a vector around a closed path and compare the final vector with the initial vector. The difference between the two tells us how much spacetime is curved in this region.

For example, we can imagine that a given vector V_α moves from A to B via two different paths:

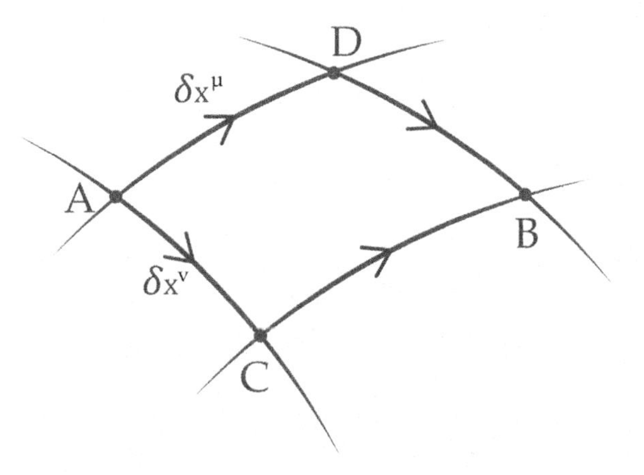

Taken together these two paths yield a closed path and we can calculate[20]

$$V_\alpha(A \to C \to B) - V_\alpha(A \to D \to B) = R_{\alpha\beta\ \mu}^{\ \ \nu} V^\beta \delta x^\mu \delta x^\nu + \dots , \tag{5.8}$$

where $R_{\alpha\beta\ \mu}^{\ \ \nu}$ denotes the corresponding (Riemann) curvature tensor[21]

$$R_{\alpha\beta\ \mu}^{\ \ \nu} = \partial_\alpha \Gamma_{\beta\ \mu}^{\ \nu} - \partial_\beta \Gamma_{\alpha\ \mu}^{\ \nu} + \Gamma_{\alpha\ \kappa}^{\ \nu} \Gamma_{\beta\ \mu}^{\ \kappa} - \Gamma_{\beta\ \kappa}^{\ \nu} \Gamma_{\alpha\ \mu}^{\ \kappa} . \tag{5.9}$$

and $\Gamma_{\alpha\ \nu}^{\ \kappa}$ are the corresponding connections.[22]

Now, what does all this have to do with gravity?

5.2.1 Gravity and Curvature Dynamics

The main idea in all previous chapters is that we can understand how interactions happen if we allow the structure of the space in question to be non-trivial. In the previous section, we saw that the structure of spacetime can be non-trivial too. And if spacetime is curved, we need the connections Γ to move vectors around consistently. In other words, if the curvature of spacetime is non-zero, the connections Γ are essential parts of the system.

Now, analogously to how we asked in the financial toy model where currency arbitrage opportunities come from, we can now ask: where does a non-zero curvature of spacetime come from?

The answer is: any object which carries non-zero energy-momentum. Thus, curvature arises from every physical object and in particular every elementary particle. And in turn, every object which carries energy-momentum is influenced by the structure of spacetime.[23]

[20] The dots here indicate further terms which do not describe the curvature of spacetime, but torsion. Torsion is another property through which a given space can have a non-trivial structure. (A space has non-zero torsion if the parallelogram that is formed by two vectors does not close.) Usually this term is neglected because, so far, no torsion of spacetime has ever been measured in experiments. This is discussed a bit further in Section 5.3. Moreover, if you wonder why we suddenly have superscript indices here, have a look at Appendix B.

[21] Since Minkowski space is four-dimensional and has a non-trivial structure, the explicit form of the curvature tensor is quite complicated. But these details are not really important for the following discussion and we therefore only quote the final result.

[22] The connections $\Gamma_{\alpha\ \nu}^{\ \kappa}$ are usually called Christoffel symbols.

[23] This is analogous to how objects carrying electric charge modify charge space and are influenced by it.

This is how gravitational interactions happen.

This means that the "charge" associated with gravity is energy-momentum. Therefore, every particle which carries energy-momentum takes part in gravitational interactions. Moreover, the curvature of spacetime $R_{\alpha\beta\ \mu}^{\ \ \nu}$ is what we call the gravitational field.[24]

[24] All this may seem quite intimidating. But take note that everything we do here is completely analogous to what we discussed already for the financial toy model and quantum mechanics. The only difference is that we are now talking about spacetime and not some internal space.

The interplay between energy-momentum and curvature is described by the Einstein equation. We can derive it by following the same steps that allowed us to derive Maxwell's equations in Section 2.1.

First of all, we need the fact that energy-momentum is conserved.[25] Mathematically, we describe energy-momentum using the energy-momentum tensor $T_{\mu\nu}$. The conservation of energy-momentum implies that there is a continuity equation[26]

$$\partial^{\mu} T_{\mu\nu} = 0. \tag{5.10}$$

[25] Using Noether's theorem this can be derived by using the fact that spacetime is homogeneous. A space is homogeneous whenever it possesses translation symmetry.

[26] This is analogous to Eq. 2.32 where, in the financial toy model, copper was conserved.

Moreover, we know already that curvature is described by the curvature tensor $R_{\alpha\beta\ \mu}^{\ \ \nu}$ (Eq. 5.9).

Attempting to connect these two equations, we observe that $T_{\mu\nu}$ has two indices while $R_{\alpha\beta\ \mu}^{\ \ \nu}$ has four. Therefore, we can't simply write

$$T_{\mu\nu} = \text{const} \times R_{\alpha\beta\ \mu}^{\ \ \nu} \,. \tag{5.11}$$

In addition, we need to make sure, since the derivative of the left-hand side is zero (Eq. 5.10), that the derivative of the right-hand side is zero, too.[27]

[27] All steps here are analogous to what we discussed in Section 2.1.

What we therefore need is a quantity, constructed from $R_{\alpha\beta\ \mu}^{\ \ \nu}$, which has only two indices and yields zero if we take the derivative. Deriving the correct mathematical object which fulfils these conditions is a somewhat complicated mathematical problem and we will therefore not dive into the details here. For our purpose, it is enough to know that it can be done and that the correct result is known as the Einstein tensor $G_{\mu\nu}$.[28]

As soon as we have this tensor, we can write down the **Einstein equation**

$$T_{\mu\nu} = CG_{\mu\nu} \,. \tag{5.12}$$

The constant C encodes how strongly spacetime reacts to the presence of energy-momentum.[29]

Now we have two indices on both sides and taking the derivative yields zero

$$\partial^{\mu} T_{\mu\nu} = 0 = \partial^{\mu} G_{\mu\nu} \,. \tag{5.13}$$

A final aspect that we need to talk about is that, once more, we need to replace our ordinary derivatives with covariant derivatives whenever there is non-zero curvature.[30] When spacetime is curved, we need to replace all derivatives with the following covariant derivative[31]

$$D_b v^a \equiv \partial_b v^a + \Gamma^a{}_{bc} v^c. \tag{5.14}$$

In other words, if we want any equation we derived so far to be valid in curved spacetime, we need to change

$$\partial_b \to D_b = \partial_b + \Gamma^a{}_{bc}. \tag{5.15}$$

By doing this, we take into account how particles react to the presence of curvature. In other words, while the Einstein

[28] For electrodynamics and our financial toy model, this task was easier because the derivative of the object we use to describe curvature ($\partial_\mu F_{\mu\nu}$) has exactly the properties we need. In contrast, the Einstein tensor can be written in terms of the curvature tensor $R_{\alpha\beta\ \mu}^{\ \ \nu}$ as follows:

$$G_{\mu\nu} = R_{\mu\nu} - \frac{1}{2} R g_{\mu\nu} \,,$$

where

$$R_{\mu\nu} = R_{\alpha\mu\ \nu}^{\ \ \alpha}$$
$$R = g^{\mu\nu} R_{\alpha\mu\ \nu}^{\ \ \alpha} \,,$$

and $g_{\mu\nu}$ denotes the metric. The concept of a metric is discussed in Appendix B.

[29] Specifically, $C = \frac{8\pi G}{c^4}$ where G is Newton's gravitational constant and c the speed of light.

[30] Reminder: this is necessary because, when we calculate the derivative, we compare the values of a function at two different locations

$$f'(a) = \lim_{h\to 0} \frac{f(a+h) - f(a)}{h}.$$

In a curved space this comparison is not as trivial as in a flat space. To move objects around consistently, we need to use the appropriate connections.

[31] This is analogous to what we discussed for curved internal spaces, e.g., around Eq. 2.55.

equation (Eq. 5.12) describes how spacetime reacts to the presence of objects carrying energy-momentum, the replacement rule in Eq. 5.15 allows us to describe how particles react to the structure of spacetime:

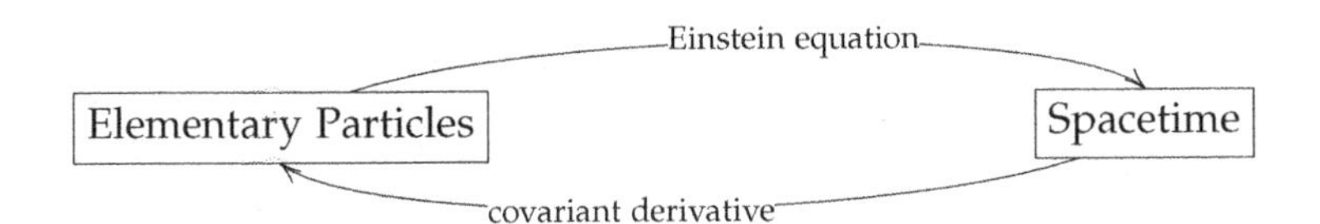

And, as mentioned before, this is how gravitational interactions happen. Objects modify spacetime and are influenced by it in turn.

5.3 Summary and a Few Loose Ends

The main point of this chapter is that we can describe gravity in a manner analogous to how we describe all other fundamental interactions. The only difference is that now it's spacetime which gets curved and not some internal space.

One loose end which we need to talk about is the charge (energy-momentum) associated with gravitational interactions. In general, we can understand the charge associated with a given space by using Noether's theorem. Noether's theorem tells us that for each symmetry of the space in question, we get a conserved quantity:[32]

[32] A conserved quantity is a quantity which does not change in time

$$\partial_t Q = 0 \rightarrow Q = \text{ const.}$$

You can find a detailed discussion and mathematical derivation of Noether's theorem in

Jakob Schwichtenberg. *Physics from Symmetry*. Springer, Cham, Switzerland, 2018b. ISBN 978-3319666303

- ▷ The conserved charge following from the $U(1)$ symmetry of charge space is what we call electric charge.
- ▷ The conserved charge following from the $SU(2)$ symmetry of isospin space is what we call isospin.

▷ The conserved charge following from the $SU(3)$ symmetry of color space is what we call color.

▷ The conserved charge following from the translational symmetry of Minkowski space is what we call energy-momentum. In particular:

- The conserved charge following from invariance under *spatial* translations $x \to x + \epsilon$ is momentum.
- The conserved charge following from invariance under *temporal* translations $t \to t + \epsilon$ is energy.

A crucial point is that since spatial components can get mixed with the time component, we need to consider them as a single (overall) conserved charge which we call energy-momentum.[33] This allows us to keep track of exactly how energy and momentum get mixed in different coordinate systems. For example, we can always achieve that the momentum of a given particle is zero by switching to a coordinate system which moves with (system attached to) the particle. Mathematically, energy-momentum is a four-vector $p_\mu = (E/c, p_1, p_2, p_3)$, where we have introduced the constant c once more to get the same units for all components.[34] Moreover, take note that the quantity we call mass m corresponds to the length of the energy-momentum four vector: $p_\mu \cdot p_\mu = E^2/c^2 - p_1^2 - p_2^2 - p_3^2 \equiv m^2c^2$.

Another interesting point is that there are additional symmetries of Minkowski space. Therefore, we get additional conserved charges. In particular:[35]

▷ The conserved charge following from the rotational symmetry of Minkowski space is what we call angular-momentum. Upon closer inspection, this conserved quantity consists of two parts: orbital angular momentum and spin. We can interpret spin as some kind of *internal*

[33] Transformations which mix spatial components with the time component are what we call boosts. A boost is a transformation to a coordinate system which moves with a constant velocity with respect to the initial coordinate system.

[34] We did the same thing when we defined the vector which describes locations in spacetime $x_\mu = (ct, x_1, x_2, x_3)$ such that

$$x_\mu \cdot x_\mu = c^2t^2 - x_1^2 - x_2^2 - x_3^2.$$

[35] Using Noether's theorem for the invariance under boosts does not yield a useful conservation law. This is discussed explicitly, for example, in my book *Physics from Symmetry*.

angular momentum.

This fact alone is not very surprising. But it becomes very interesting once we recall the roles of all charges we discussed so far.

- ▷ Particles carrying electric charge change the structure of charge space.
- ▷ Particles carrying isospin change the structure of isospin space.
- ▷ Particles carrying color change the structure of color space.
- ▷ Particles carrying energy-momentum change the structure of spacetime.

What then is the role of angular momentum?

While one can argue that orbital angular momentum is somewhat different from the other charges since it involves the (arbitrary) choice of an origin in a coordinate system about which rotational motion is measured, spin is a property which is carried by individual elementary particles. In other words, analogous to how some particles carry electric charge, there are particles which carry spin.[36]

From the perspective described in this book it seems natural that the presence of spin at some location has an effect on the geometry of the arena of physics, too.

One possibility is that spin leads to a **torsion** of spacetime while energy-momentum causes a curvature.[37] In Einstein's general relativity we simply assume that the torsion of spacetime is zero. The model we end up with if we relax this assumption is known as **Einstein-Cartan gravity**.[38]

[36] In fact, all known elementary particles carry spin except for the recently discovered Higgs boson.

[37] Torsion was already mentioned in the text around Eq. 5.8. If we calculate how a vector changes if we move it between two points via two different paths, we find that in general, the answer consists of two terms. The first term contains information about curvature, the second term about torsion. In other words, there are two possible reasons why its final state and initial state are different if we move a vector along a closed path. One reason is that the space we are moving in is curved, the second one is that the space has non-zero torsion.

[38] "*André Weil recalled that he was present at a reception for Einstein by Romain Roland; and that at the reception he was present during a conversation between Eli Cartan and Einstein when Cartan asked Einstein why he had not allowed for torsion in his theory; and André felt even then that Einstein did no really understand what Cartan was talking about!*" [Chandrasekhar, 2010].

However, so far, this has never been confirmed experimentally. In other words, a non-zero torsion of spacetime has never been measured.[39]

[39] This does not mean that Einstein-Cartan gravity is wrong. Maybe the effect of spin on the geometry of spacetime is simply too small to be measurable with present day technology. It is worth mentioning in this context that the effect of individual elementary particles on the curvature of spacetime has never been measured either because the energy-momentum they carry is simply too small. But the effect of huge macroscopic objects on the geometry of spacetime has, of course, been observed in experiments.

6

Closing Words

If at this point you have lots of questions, I think that this book has fulfilled its purpose.

How do we use Maxwell's equations in electrodynamics? How can we use Schrödinger's equation to describe quantum systems? How do we calculate the probability that certain scattering processes happen? How can we solve Einstein's equation and use it to describe our universe?

Of course, the answer to any of these questions is far beyond the scope of this book. But I think you are now optimally prepared to dive deeper.

Before I recommend some of my favorite books which deal specifically with these questions, I want to put things into context. Hopefully, this will allow you to see clearly which theory you should focus on if you want to answer a specific question.

Our theories make up, in some sense, a (2×2) matrix. There are theories in which we focus on fields and there are theories in which particles are our main focus.[1] In addition, we can describe both particles and fields in either a classical or quantum framework. This yields, in total, four different combinations:

[1] Take note that it's possible to describe particles in quantum field theory, but they are understood as excitations of the underlying quantum fields.

	Classical	Quantum
Particles	Classical Mechanics	Quantum Mechanics
Fields	Classical Field Theory	Quantum Field Theory

It's crucial to understand that no single framework is *the best*. Each of them is useful and which one we use always depends on the system we want to describe.

▷ Field theories are great for describing the fundamental interactions themselves and systems in which the particle number is not conserved.

▷ In contrast, particle theories are great for describing individual objects.

▷ Classical theories are still the best tools we have to describe nature at macroscopic scales.

▷ But we need quantum theories to describe the behavior of microscopic systems.

To understand this a bit better, let's talk about concrete examples:[2]

[2] As mentioned above already, there are no macroscopic models describing the weak field (isospin space curvature) and the strong field (color space curvature) because the associated interactions are extremely short-ranged.

▷ We describe the dynamics of the electromagnetic field (charge space curvature $F_{\mu\nu}$) and gravitational field (spacetime curvature $R_{\alpha\beta}{}^{\nu}{}_{\mu}$) on *macroscopic* scales using classical field theory. The resulting models are known as classical electrodynamics and general relativity.

- ▷ We describe the effects of interaction fields like $F_{\mu\nu}$ on individual *macroscopic* objects using classical mechanics.
- ▷ We describe the effects of interaction fields like $F_{\mu\nu}$ on individual *microscopic* objects using quantum mechanics.
- ▷ We describe the dynamics of interaction fields on *microscopic* scales using quantum field theory. For example, the dynamics of the electromagnetic field $F_{\mu\nu}$ on microscopic scales is described by quantum electrodynamics and the dynamics of the strong field $G_{\mu\nu}$ is described by quantum chromodynamics.[3]

[3] Take note that so far, there is no generally accepted model which describes the gravitational field on microscopic scales, i.e. gravitational interactions between elementary particles. This is known as the problem of quantum gravity and we will talk a bit more about this below.

An important aspect of quantum field theory is that we can also use it to describe the effects of the interaction fields on elementary particles. Therefore, we can use it directly to describe how elementary particles interact. In contrast, we need both classical field theory and classical mechanics to describe electromagnetic interactions between macroscopic objects for example. In particular, we need classical field theory to describe the electromagnetic field and classical mechanics to describe its effect on macroscopic objects.

But this does not mean that quantum field theory renders all other frameworks obsolete. For example, we can't use quantum field theory to describe gravitational interactions between planets. Therefore, we really need all frameworks to describe nature at all scales.

Now, what are some good sources to learn more about these theories?

6.1 Reading Recommendations

Great books to learn the basic ideas of classical mechanics are

▷ **The Lazy Universe** by Jennifer Coopersmith[4].

▷ **Introduction to Classical Mechanics** by David Morin[5].

Moreover, you might enjoy my book

▷ **No-Nonsense Classical Mechanics**[6].

Instead of learning classical field theory in general, most students learn classical electrodynamics and general relativity directly. My favorite student-friendly electrodynamics textbooks are

▷ **Volume 2 of the Feynman Lectures**[7]

▷ **A student's guide to Maxwell's equations** by Daniel A. Fleisch[8]

▷ **Introduction to Electrodynamics** by David J. Griffiths[9]

You also might enjoy my book

▷ **No-Nonsense Electrodynamics**[10].

[4] Jennifer Coopersmith. *The lazy universe.* Oxford University Press, 2017. ISBN 9780198743040

[5] David Morin. *Introduction to classical mechanics: with problems and solutions.* Cambridge University Press, 2008. ISBN 9780511808951

[6] Jakob Schwichtenberg. *No-Nonsense Classical Mechanics.* No-Nonsense Books, 2019. ISBN 978-1096195382

[7] Richard Feynman. *The Feynman lectures on physics.* Addison-Wesley, San Francisco, Calif. Harlow, 2011. ISBN 9780805390650

[8] Daniel Fleisch. *A student's guide to Maxwell's equations.* Cambridge University Press, 2008. ISBN 978-0521701471

[9] David Griffiths. *Introduction to electrodynamics.* Pearson Education Limited, Harlow, 2014. ISBN 9781292021423

[10] Jakob Schwichtenberg. *No-Nonsense Electrodynamics.* No-Nonsense Books, 2018a. ISBN 978-1790842117

Great books to start learning more about Einstein's theory of general relativity are

- ▷ **Relativity, Gravitation and Cosmology** by Ta-Pei Cheng[11].
- ▷ **Einstein Gravity in a Nutshell** Anthony Zee[12].

Two illuminating books to learn how quantum mechanics works in practice are

- ▷ **Lectures on Physics Volume 3** by Richard Feynman[13].
- ▷ **Quantum Mechanics** by David J. Griffiths[14].

In addition, since most textbooks make quantum mechanics appear much more complicated than it really is, I decided to write one called

- ▷ **No-Nonsense Quantum Mechanics**[15].

The best book to get started with quantum field theory is

- ▷ **Student Friendly Quantum Field Theory** by Robert D. Klauber[16].

[11] Ta-Pei Cheng. *Relativity, Gravitation and Cosmology: A Basic Introduction*. Oxford University Press, 2nd edition, 1 2010. ISBN 9780199573646

[12] Anthony Zee. *Einstein Gravity in a Nutshell*. Princeton University Press, 1st edition, 5 2013. ISBN 9780691145587

[13] Richard Feynman. *The Feynman lectures on physics*. Addison-Wesley, San Francisco, Calif. Harlow, 2011. ISBN 9780805390650

[14] David Griffiths. *Introduction to quantum mechanics*. Pearson Prentice Hall, 2005. ISBN 9780131118928

[15] Jakob Schwichtenberg. *No-Nonsense Quantum Mechanics*. No-Nonsense Books, 2018c. ISBN 978-1719838719

[16] Robert D. Klauber. *Student Friendly Quantum Field Theory*. Sandtrove Press, 2nd edition, 12 2013. ISBN 9780984513956

Moreover, the keyword you should look out for if you want to learn more about the geometrical perspective described in this book is "fiber bundle".[17]

[17] For a few more details on fiber bundles, see Appendix D.

Good starting points to learn more about fiber bundles are

▷ **Fiber Bundles and Quantum Theory** by Herbert J. Bernstein and Anthony V. Phillips[18].

▷ **Gauge fields, knots, and gravity** by John Baez and Javier P. Muniain[19].

▷ **The Road to Reality** by Roger Penrose[20].

A great mathematics textbook on fiber bundles and how they are used in physics is

▷ **Topology, Geometry, and Gauge Fields: Foundations** by Gregory Naber[21].

A particularly illuminating example of a non-trivial fiber bundle is the Hopf bundle. A great discussion of the Hopf bundle and why it is important in physics can be found in

▷ **Some Elementary Gauge Theory Concepts** by Hong-Mo Chan, Sheung Tsun Tsou[22].

[18] H. J. Bernstein and A. V. Phillips. Fiber Bundles and Quantum Theory. *Sci. Am.*, 245: 94–109, 1981

[19] John Baez and Javier P. Muniain. *Gauge fields, knots, and gravity*. World Scientific, Singapore River Edge, N.J, 1994. ISBN 9789810217297

[20] Roger Penrose. *The road to reality: a complete guide to the laws of the universe*. Vintage Books, New York, 2007. ISBN 9780679776314

[21] Gregory Naber. *Topology, geometry and Gauge fields: foundations*. Springer Science+Business Media, LLC, New York, 2011. ISBN 9781441972538

[22] Chan Hong-Mo and Tsou Sheung Tsun. *Some elementary gauge theory concepts*. World Scientific, Singapore River Edge, N.J, 1993. ISBN 9789810210809

Last but not least, a fantastic book to learn more about how fiber bundles can be used to describe the financial market is

▷ **The Physics of Finance** by Kirill Ilinski[23].

[23] Kirill Ilinski. *Physics of finance: gauge modelling in non-equilibrium pricing*. John Wiley, Chichester England New York, 2001. ISBN 9780471877387

In contrast to the modest goal of this book, Ilinski tries to construct a realistic model of the financial market.

6.2 Outlook

The main idea described in this book is that we can understand all fundamental interactions beautifully from a common perspective. Therefore, it may seem as if physics were almost a finished project. But this couldn't be further from the truth.

There are still *lots* of open questions.[24]

[24] Several open questions have already been mentioned in passing.

Most importantly, there is still no generally accepted way to incorporate general relativity in the framework of quantum field theory. While Einstein's classical field model works perfectly at large scales, we know that at small scales quantum field theory holds sway.[25] But there are lots of other more modest open questions, for example, about the role of torsion in general relativity.[26]

[25] Take note that as long as we stick to low energies, we can incorporate General Relativity as an *effective* quantum field model. But we know that this theory does not remain valid at high-energy scales. In other words, we know that this effective model must be replaced with a better model at high energies. (That's the whole point of effective models. But from a modern perspective, all the models we have come up with so far are effective models which are only valid up to some cutoff scale.)

[26] This problem was mentioned already at the end of the last chapter.

Moreover, there are lots of open questions about the particle zoo. Why exactly these particles? Why do they carry the charges they do? In addition, there are good reasons to believe that there are additional elementary particles, for example, to explain the masses of neutrinos and the observed dark matter density.

I personally think there is still much to come and maybe a completely new framework is needed to overcome the present obstacles. Future developments will be extremely interesting and I hope you will continue to follow the story and maybe contribute something yourself.

One Last Thing

It's impossible to overstate how important reviews are for an author. Most book sales, at least for books without a marketing budget, come from people who find books through the recommendations on Amazon. Your review helps Amazon figure out what types of people would like my book and makes sure it's shown in the recommended products.

I'd never ask anyone to rate my book higher than they think it deserves, but if you like my book, please take the time to write a short review and rate it on Amazon. This is the biggest thing you can do to support me as a writer.

Each review has an impact on how many people will read my book and, of course, I'm always happy to learn about what people think about my writing.

PS: If you write a review, I would appreciate a short email with a link to it or a screenshot to Jakobschwich@gmail.com. This helps me to take note of new reviews. And, of course, feel free to add any comments or feedback that you don't want to share publicly.

Part V
Appendices

A

Statistics

A.1 Probability Distribution

A probability distribution tells us the probability of each possible outcome. Formulated differently, a probability distribution eats a possible outcome o and spits out the corresponding probability $\chi(o)$.

For example, the probability distribution for a regular die is[1]

$$\chi(1) = 1/6$$
$$\chi(2) = 1/6$$
$$\chi(3) = 1/6$$
$$\chi(4) = 1/6$$
$$\chi(5) = 1/6$$
$$\chi(6) = 1/6.$$

But we can also imagine that we are dealing with an oddly

[1] An important property is that the probabilities add to 1 = 100 %. This is necessarily the case because one of the outcomes must happen and a total probability of more than 100% doesn't make sense.

shaped die. Thus, our probability could look as follows[2]

[2] Again, take note that the probabilities add to 1.

$$\chi(1) = 1/2$$
$$\chi(2) = 1/6$$
$$\chi(3) = 1/12$$
$$\chi(4) = 1/12$$
$$\chi(5) = 1/12$$
$$\chi(6) = 1/12.$$

In words, this means that with this die, we are much more likely to get a 1 than, say, a 6.

A.2 Mean

One of the simplest but at the same time most important statistical tools is the so-called **mean**[3].

[3] Alternative names for the mean are: *expected value, expectation value* or simply the *average*.

The mean of a quantity is the average value we obtain when we repeat a given experiment many times. We use it whenever we are forced to make probabilistic predictions.[4]

[4] Think: flipping a coin, or tossing a die.

For example, imagine the following situation: A friend offers to play a game. She flips a coin. If it lands on tails, she pays you $1.5. But if it lands on heads, you have to pay her $1. By calculating the expectation value for the outcome, you can decide whether or not you should play this game.

▷ The probability that a coin lands on heads is $p_1 = 50\%$. In this case, the outcome for you is: $x_1 = -\$1$

▷ Equally, the probability that a coin lands on tails is $p_2 = 50\%$. In this case the outcome for you is: $x_2 = +\$1.5$

The mean is defined as the sum over each outcome times

the probability of the outcome

$$\begin{aligned} \text{mean} &= \sum_i x_i P_i = x_1 P_1 + x_2 P_2 \\ &= -\$1 \times 50\% + \$1.5 \times 50\% = \$0.25\,. \end{aligned} \tag{A.1}$$

So, if you play this game many times, you will make a profit. On average, you make \$0.25 per game. For example, let's say you play the game only twice: you win the first time and lose the second time. For the win in the first game, you get \$1.5. For the loss in the second game, you lose \$1. In total, you therefore win \$0.5 in two games. This equals \$0.25 per game.

Of course, the system here is so simple that we could have guessed this without calculating anything. But for more complicated systems the situation quickly gets messy.

To summarize: the mean is the average of the outcomes if we play the game many, many times. A common mathematical notation for the mean of a quantity x looks like this: $\langle x \rangle$.

A.3 Standard Deviation

There is one other statistical notion that we need whenever we describe a system in probabilistic terms. It gives us information on how much the system deviates from the most probable outcome *on average*. In other words, this statistical quantity gives an indication of how much our measurements are spread out. If it is zero, the outcome is the same for every measurement; we measure the same value all the time. If it is non-zero, it's possible to measure different values.

The statistical quantity I'm talking about is called the **stan-**

dard deviation. It tells us how much our measurements (on average) get spread around the average outcome (= the expectation value):[5]

$$\Delta x = \sqrt{\langle x^2 \rangle - \langle x \rangle^2}\,. \tag{A.2}$$

[5] This strange looking definition will make sense in a moment. And an aside: The same definition without the square root is known as **variance**.

The second term under the square root is simply the expectation value squared. For the first term, we calculate an expectation value again, but this time we square each possible outcome before we weigh it with the corresponding probability

$$\langle x^2 \rangle = \sum_i x_i^2 P_i\,. \tag{A.3}$$

Compare this with

$$\langle x \rangle^2 = \left(\sum_i x_i P_i\right)^2. \tag{A.4}$$

Yes! It makes a difference whether we square the weighted sum of the outcomes or we sum the weighted squares of the outcomes![6] To see concretely the meaning of the standard deviation, let's consider the example from above again. The standard deviation reads

[6] For example, for $x = \{3, -5, 9\}$ we have

$$\sum_i x_i^2 = (3)^2 + (-5)^2 + (9)^2 = 115$$

whereas

$$\left(\sum_i x_i\right)^2 = (3-5+9)^2 = 49\,.$$

$$\begin{aligned}
\Delta x &= \sqrt{\langle x^2 \rangle - \langle x \rangle^2} \\
&\quad \text{Eq. A.3 and Eq. A.4} \\
&= \sqrt{\sum_i x_i^2 P_i - \left(\sum_i x_i P_i\right)^2} \\
&\quad \text{Eq. A.1} \\
&= \sqrt{\sum_i x_i^2 P_i - (\$0.25)^2} \\
&= \sqrt{((-\$1)^2 \times 0.5 + (\$1.5)^2 \times 0.5) - (\$0.25)^2} \\
&= \$\sqrt{1.625 - 0.0625} \approx \$1.25.
\end{aligned}$$

There are two important things to take note of:

- ▷ The result isn't zero. It makes a difference whether we square each term in the sum or the complete sum. We can see this explicitly here. Otherwise, the standard deviation would always be zero.
- ▷ If we only knew that the expectation value of some game is \$0.25 we would be missing a lot of information. Naively, we could think that, since \$0.25 is such a small amount of money, the bets involved are not high at all. However, imagine that we make our example more extreme by proposing that if the coin lands on heads you have to pay \$10000, and if it lands on tails you get \$10000.5. The expectation value is again $\langle x \rangle = 0.25$. But the game is much more extreme now. I wouldn't like to play it. While the odds are still in your favor, now there is also a high risk for you to lose *a lot* of money. This kind of information is encoded in the standard deviation. For this modified game the standard deviation is $\Delta x \approx \$10000.25$. We can therefore see immediately that a lot of money is at stake.

One final comment on the idea behind the definition of the standard deviation: The logic behind squaring each possible outcome in the first term on the right-hand side is to avoid that terms cancel because they have opposite signs. This is exactly what happens when we calculate the expectation value. In the coin example, if the outcome was heads we had to pay \$1, which mathematically means $x_1 = -\$1$. Thus, in the formula for the mean, the two terms almost canceled since they have different signs. But now we want to get information about the spread in our measurements. Hence, we need a notion that takes the *absolute size* of the outcomes into account. We accomplish this by squaring each outcome and only then multiplying each term calculated this way with the corresponding probability. We then also square the expectation value such that it has the same units as the thing we just calculated. Otherwise, we would be comparing apples with pears.[7]

[7] In the example $\$^2$ with \$.

B

Metrics

In this appendix, we talk about Eq. 5.4 a bit more systematically.

In general, the mathematical tool which allows us to calculate the distance between two points in a given space is called the **metric**. In general, a metric encodes important information about the structure of the space.

For example, the mathematical relations to calculate the distance between two points A and B on a Euclidean plane and between two points C and D on a sphere are different. In the first case, the distance is given by the familiar Pythagorean formula, irrespective of the actual distance between the points, while in the second case, that formula is only (approximately) valid for points which are close to each other on the surface of the sphere. It is the metric that encodes how distance between points can be calculated in a given space.

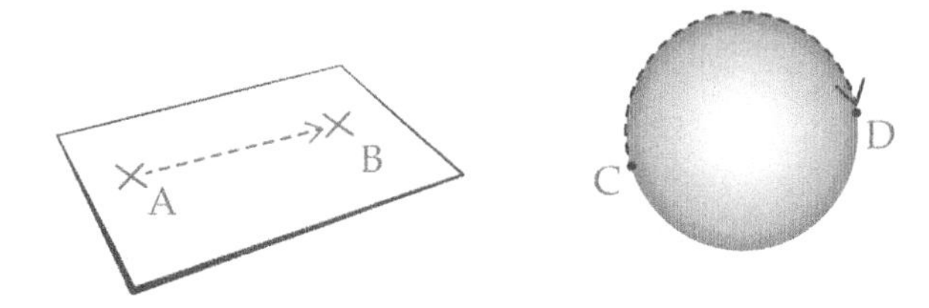

The best way to understand what a metric is, is to consider concrete examples. We can calculate the distance between two points

$$A = \begin{pmatrix} x_0 \\ x_1 \\ x_2 \\ x_3 \end{pmatrix}, \quad B = \begin{pmatrix} \tilde{x}_0 \\ \tilde{x}_1 \\ \tilde{x}_2 \\ \tilde{x}_3 \end{pmatrix} \tag{B.1}$$

by calculating the length of the vector

$$\vec{v} \equiv \begin{pmatrix} \Delta x_0 \\ \Delta x_1 \\ \Delta x_2 \\ \Delta x_3 \end{pmatrix} = \begin{pmatrix} \tilde{x}_0 - x_0 \\ \tilde{x}_1 - x_1 \\ \tilde{x}_2 - x_2 \\ \tilde{x}_3 - x_3 \end{pmatrix} \tag{B.2}$$

which connects them. In a four-dimensional Euclidean space, we find[1]

[1] Technically, we are calculating the scalar product of the vector with itself.

$$d(A,B) = \vec{v} \cdot \vec{v} = \Delta x_0^2 + \Delta x_1^2 + \Delta x_2^2 + \Delta x_3^2 . \tag{B.3}$$

But in Minkowski space, we find

$$d(A,B) = \vec{v} \cdot \vec{v} = \Delta x_0^2 - \Delta x_1^2 - \Delta x_2^2 - \Delta x_3^2 . \tag{B.4}$$

We describe this fact by saying that a Euclidean and a Minkowski space have a different metric g and write the scalar product as $\vec{v}^T g \vec{v}$.

Now, for

$$g_E = \begin{pmatrix} 1 & 0 & 0 & 0 \\ 0 & 1 & 0 & 0 \\ 0 & 0 & 1 & 0 \\ 0 & 0 & 0 & 1 \end{pmatrix} \tag{B.5}$$

we find the correct scalar product of a Euclidean space

$$\vec{v}^T g_E \vec{v} = \begin{pmatrix} \Delta x_0 & \Delta x_1 & \Delta x_2 & \Delta x_3 \end{pmatrix} \begin{pmatrix} 1 & 0 & 0 & 0 \\ 0 & 1 & 0 & 0 \\ 0 & 0 & 1 & 0 \\ 0 & 0 & 0 & 1 \end{pmatrix} \begin{pmatrix} \Delta x_0 \\ \Delta x_1 \\ \Delta x_2 \\ \Delta x_3 \end{pmatrix}$$

$$= \Delta x_0^2 + \Delta x_1^2 + \Delta x_2^2 + \Delta x_3^2 . \tag{B.6}$$

And for

$$g_M = \begin{pmatrix} 1 & 0 & 0 & 0 \\ 0 & -1 & 0 & 0 \\ 0 & 0 & -1 & 0 \\ 0 & 0 & 0 & -1 \end{pmatrix} \tag{B.7}$$

we find the correct scalar product of Minkowski space[2]

$$\vec{v}^T g_M \vec{v} = \begin{pmatrix} \Delta x_0 & \Delta x_1 & \Delta x_2 & \Delta x_3 \end{pmatrix} \begin{pmatrix} 1 & 0 & 0 & 0 \\ 0 & -1 & 0 & 0 \\ 0 & 0 & -1 & 0 \\ 0 & 0 & 0 & -1 \end{pmatrix} \begin{pmatrix} \Delta x_0 \\ \Delta x_1 \\ \Delta x_2 \\ \Delta x_3 \end{pmatrix}$$

$$= \Delta x_0^2 - \Delta x_1^2 - \Delta x_2^2 - \Delta x_3^2 . \tag{B.8}$$

We call g_E the **Euclidean metric** and g_M the **Minkowski metric**.[3]

In index notation, we can write the scalar product in Minkowski space as:

$$(\Delta s)^2 = v_\mu \eta^{\mu\nu} v_\nu$$

$$= \begin{pmatrix} \Delta x_0 & \Delta x_1 & \Delta x_2 & \Delta x_3 \end{pmatrix} \begin{pmatrix} 1 & 0 & 0 & 0 \\ 0 & -1 & 0 & 0 \\ 0 & 0 & -1 & 0 \\ 0 & 0 & 0 & -1 \end{pmatrix} \begin{pmatrix} \Delta x_0 \\ \Delta x_1 \\ \Delta x_2 \\ \Delta x_3 \end{pmatrix}$$

$$= \Delta x_0^2 - \Delta x_1^2 - \Delta x_2^2 - \Delta x_3^2 \tag{B.9}$$

Moreover, it is conventional to introduce superscript indices

[2] This is equivalent to Eq. 5.4. The only difference is that now we use a different vector and don't interpret the zeroth component as time.

[3] Take note that it is conventional to use the symbol η for the Minkowski metric.

in order to avoid writing the Minkowski metric all the time:

$$x^\mu \equiv \eta^{\mu\nu} x_\nu \tag{B.10}$$

or equally

$$\begin{aligned} y^\nu &\equiv \eta^{\nu\mu} y_\mu \\ &= \eta^{\mu\nu} y_\mu \,. \end{aligned} \tag{B.11}$$

↷ the Minkowski metric is symmetric $\eta^{\mu\nu} = \eta^{\nu\mu}$

This allows us to write the scalar product as follows:

$$x \cdot y \equiv x_\mu \eta^{\mu\nu} y_\nu = x_\mu y^\mu = x^\nu y_\nu . \tag{B.12}$$

The bottom line is that when you see a superscript index in physics, you should remember that this is usually a shorthand notation for the Minkowski metric.

C

Group Theory

In Section 1.1, we talked about the general definition of a symmetry. Intuitively, you can imagine that someone holds the object in question in front of you. If you close your eyes while the person performs a transformation and afterwards you can't tell if the person has changed anything at all, the transformation is a symmetry of the object.

The branch of mathematics which deals with symmetries is called **group theory**. A group is a set of transformations which fulfill special rules plus an operation that tells us how to combine the transformations.[1]

[1] The rules are known as group axioms and they can be motivated by investigating an intuitive symmetry like rotational symmetry, see `www.jakobschwichtenberg.com/group-axioms`

A square consists of an infinity of points. A symmetry of the square is a transformation which maps these points onto themselves.

Some rotations about the origin (not all) are symmetries of a square. This means that certain rotations map each point that our square consists of onto another point on the square. Therefore, we can say that the set of points our square consists of is invariant under such a transformation.

To understand this, let's focus on one particular corner of the square. If we rotate the square by 50°, we end up with a point outside the original set.

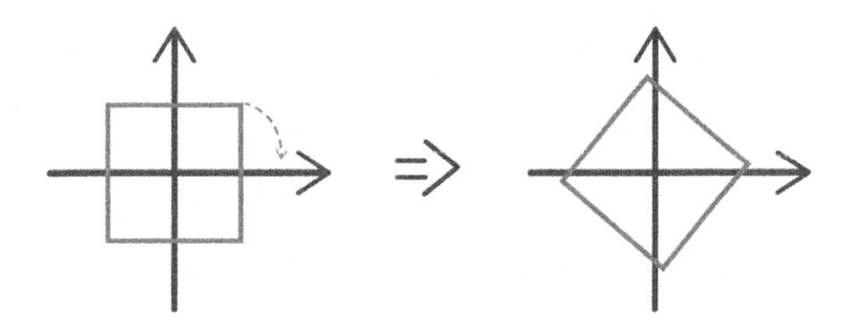

But if we rotate the square by 90° we find that the corner point gets mapped to a location where another corner point was previously located.

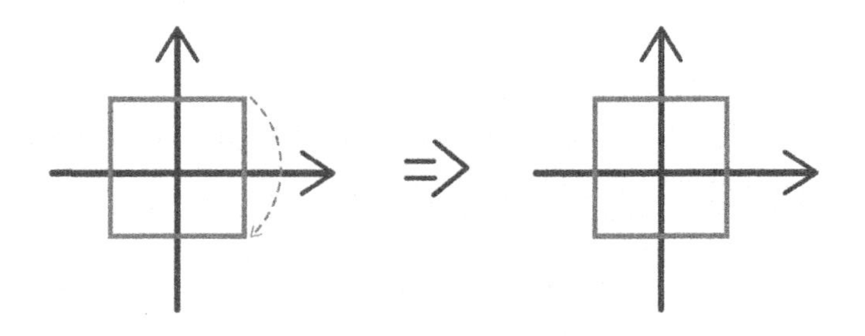

Therefore, rotations by 90°, 180°, 270° and of course 0° are symmetries of a square, while all other rotations aren't.[2] All these rotations taken together plus a rule which allows us to combine rotations are the symmetry group of the square.

[2] Take note that four additional symmetries of a square are mirror symmetries; namely with respect to the horizontal, vertical, and (two) diagonal axes.

90° rotation
combination rule
0° rotation
+
180° rotation =
270° rotation
symmetry group of a square

Since the transformation parameter (rotation angle) can only take on a discrete set of values, we say the group is discrete.

The situation is somewhat different when we investigate transformations that leave a circle unchanged. While a circle also consists of lots of points, there are many more transformations which map these points onto themselves.

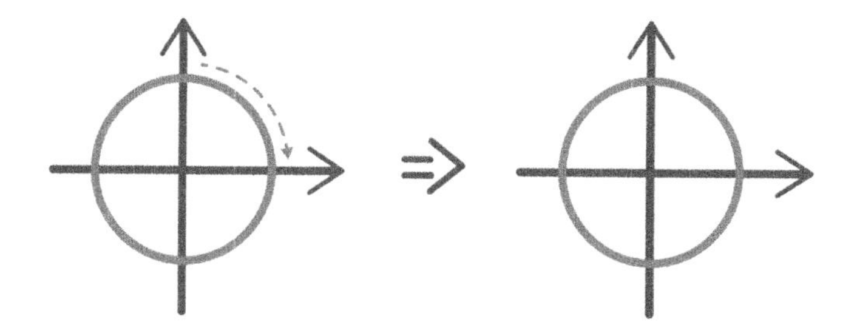

Concretely, a circle is invariant under *all* rotations about the origin. Again, our symmetry group consists of transformations and an additional rule which allows us to combine the group elements in such a way that we get another group element. For example, a rotation by 89.2° plus a rotation by 90°, yields a rotation by 179.2° which is another symmetry of the circle.

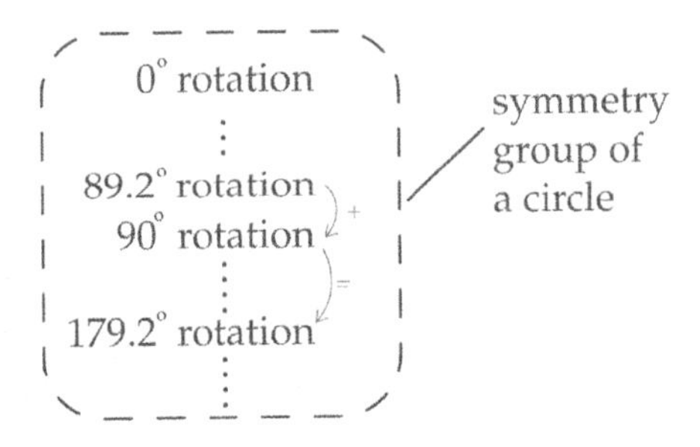

Since the transformation parameter (rotation angle) can take on arbitrary continuous values, we say the group is continuous.

The main idea of group theory is to distill the defining features of such symmetries into a precise mathematical form. This yields the axioms of group theory. Afterwards, all kinds of structures which fulfill these axioms can be investigated. This is a useful approach because there exist more abstract symmetries (e.g. of equations) and group theory allows us to apply the basic lessons we learned by studying intuitive examples.

C.1 Examples

In the previous chapters, we learned that we can make sense of the internal space associated with a given theory by using symmetries. In our financial toy model the symmetry group is the **dilation group** which consists of all possible dilations

$$f = e^{\epsilon}, \qquad \text{with} \qquad \epsilon \in \mathbb{R}. \tag{C.1}$$

of the given currency. The mathematical name for this group is $GL^+(1, R)$, the one-dimensional real general linear group with positive determinant.

In quantum mechanics the symmetry group is the unitary group of dimension one, denoted by $U(1)$, and consists of all possible phase shifts

$$f = e^{i\epsilon}, \qquad \text{with} \qquad \epsilon \in \mathbb{R}. \tag{C.2}$$

The term unitary refers to the fact that these phase shifts fulfill the condition $(e^{i\epsilon})^\star e^{i\epsilon} = 1$. The difference between $GL^+(1,R)$ and $U(1)$ is the factor i in the exponent of the transformation operators.

An extremely interesting aspect of groups is that we can understand them geometrically.

C.2 Geometry of Symmetries

First of all, let's recall two observations that we made in previous chapters.

▷ In Section 1.2, we noticed that our financial market has a symmetry because we can arbitrarily shift all the local prices. In mathematical terms, this means that we can multiply all prices by any number without actually changing anything to the dynamics of the system.[3] The set of symmetry transformations therefore consists of all possible multiplications with real numbers. Geometrically, these numbers yield a line.

▷ Moreover, we learned that the dynamics of quantum systems is unaffected if we multiply the wave function ψ by a unit complex number $e^{i\epsilon}$. Mathematically, these lie on the unit circle in the complex plane.

These are examples of a beautiful and deep insight: we can understand symmetry transformations geometrically.[4]

Geometrically, the dilation group is a line.

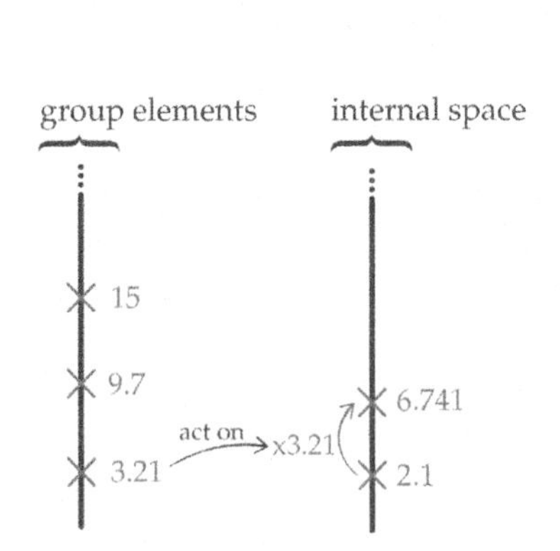

[3] However, we need to multiply all prices with the same number.

[4] To be more precise: Lie groups are mathematically differentiable manifolds. In the following, whenever the word "group" is used, I mean a Lie group. Lie groups describe continuous symmetries, e.g. the symmetries of a circle. Continuous symmetries can be parameterized by a continuous parameter like the angle of rotation θ. There are also discrete groups which describe, for example, the symmetries of a square. (Only rotations by $0°, 90°, 180°, 270°$ are symmetries of a square and therefore, these transformations can not be parameterized by a continuous parameter.)

The group $U(1)$ is geometrically a circle.[5]

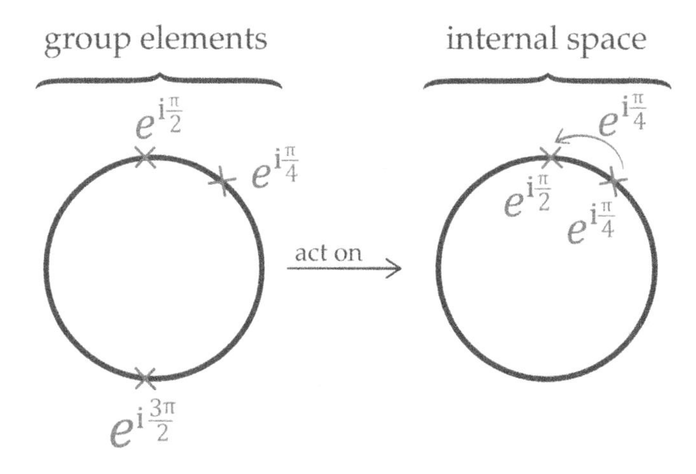

[5] Recall that this follows from Euler's formula

$$z = e^{i\phi}$$
$$= \cos(\phi) + i\sin(\phi)$$
$$\equiv \mathrm{Re}(z) + i\mathrm{Im}(z).$$

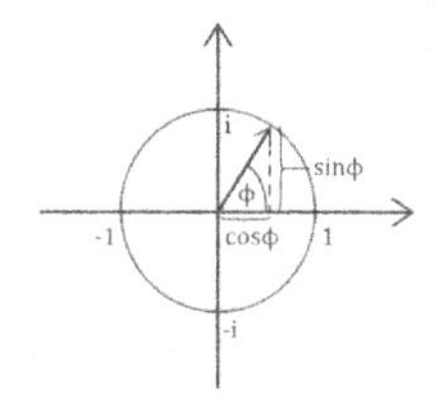

Concretely, this means

$$e^{i\frac{\pi}{2}} = \cos(\pi/2) + i\sin(\pi/2) = 0 + 1 \cdot i = i$$
$$e^{i\pi} = \cos(\pi) + i\sin(\pi) = -1 + 0 \cdot i = -1$$
$$e^{i\frac{3\pi}{2}} = \cos(3\pi/2) + i\sin(3\pi/2) = 0 - 1 \cdot i = -i$$
$$e^{i2\pi} = \cos(2\pi) + i\sin(2\pi) = 1 + 0 \cdot i = 1$$

Moreover, take note that

$$e^{i\frac{\pi}{4}} e^{i\frac{\pi}{4}} = e^{i\frac{\pi}{4} + i\frac{\pi}{4}} = e^{i\frac{\pi}{2}}$$

In the following appendix (Appendix D) we will use this idea to understand the arenas we use to describe our systems in geometrical terms.

C.3 Lie Algebras

There is one property that makes continuous symmetries (Lie groups) especially nice to deal with: they have elements which are arbitrarily close to the identity transformation[6]. A discrete group has no element that is arbitrarily close to the identity.

[6] The identity transformation is the transformation that does nothing at all. For example, for rotations, the identity transformation is a rotation by 0°.

Take, for example, the symmetries of a square. The set of transformations that leaves the square invariant consists of four rotations: a rotation by 0°, by 90°, by 180° and one by 270°, plus some mirror symmetries. A rotation by 0,000001°, which is very close to the identity transformation (a rotation by 0°), is not a symmetry.

Next, take a look at the symmetries of a circle. Certainly, a rotation by 0,000001° is a symmetry of the circle. Mathematically, we write an element g close to the identity I like this:

$$g(\epsilon) = I + \epsilon G \tag{C.3}$$

where ϵ (epsilon) is, as usual in mathematics, some really, really small number and G is an object, called a **generator**, which we will talk about in a moment.

Such small transformations barely change anything when acting on an object. In the smallest possible case, such transformations are called **infinitesimal transformations**. However, when we repeat such an infinitesimal transformation many times, we end up with a finite transformation. Think about rotations: Many small rotations in one direction are equivalent to one large rotation in the same direction.

Mathematically, we can write the idea of repeating a small transformation many times as follows

$$h(\theta) = (I + \epsilon G)(I + \epsilon G)(I + \epsilon G)... = (I + \epsilon G)^k, \tag{C.4}$$

where k denotes how often we repeat the small transformation.

If θ denotes some finite transformation parameter, e.g. 50° or so, and N is some huge number that makes sure we are close to the identity, we can write Eq. (C.3) as

$$g(\epsilon) = g\left(\frac{\theta}{N}\right) = I + \frac{\theta}{N}G\,. \tag{C.5}$$

The transformations we want to consider are the smallest possible, which means N must be the largest possible number, i.e., $N \to \infty$. To get a finite transformation from such an infinitesimal transformation, one has to repeat the infinitesimal transformation infinitely often. Mathematically

$$h(\theta) = \lim_{N\to\infty} (I + \frac{\theta}{N} G)^N, \qquad \text{(C.6)}$$

which, in the limit $N \to \infty$, is the exponential function[7]

$$h(\theta) = \lim_{N\to\infty} (I + \frac{\theta}{N} G)^N = e^{\theta G}. \qquad \text{(C.7)}$$

[7] This is one possible expression of the exponential function. We derive another (equivalent) expression in terms of an infinite series in Appendix E.

The bottom line is that the object G *generates* the finite transformation h. This is why we call such objects **generators**.

While a Lie group consist of group elements, we say that the corresponding generators live in a **Lie algebra** which belongs to the group. A main feature of the Lie algebra is that we need a different rule to combine its elements. While the group elements can be combined using the group product (e.g. matrix multiplication), the Lie algebra product is somewhat more complicated and involves a so-called Lie bracket. The most famous Lie bracket is the commutator

$$[A, B] = AB - BA.$$

So the natural product of two generators reads

$$A \circ B = AB - BA \equiv [A, B].$$

We need such a weird looking product rule because neither the product AB nor BA are necessarily generators of the group in question. However, $AB - BA$ is always another generator and therefore exactly what we need.[8]

[8] You can find a proof of this fact in John Stillwell. *Naive Lie Theory*. Springer, 1st edition, August 2008. ISBN 978-0387782140

We can understand this by using the explicit product of two group elements and by rewriting it using the corresponding generators:

$$\underbrace{g}_{\in G} \circ \underbrace{h}_{\in G} = \underbrace{e^{X+Y+\frac{1}{2}[X,Y]+\frac{1}{12}[X,[X,Y]]-\frac{1}{12}[Y,[X,Y]]+\ldots}}_{\in G} \qquad \text{(C.8)}$$

To get the formula on the right-hand side, we use the famous Baker-Campbell-Hausdorff formula.[9]

[9] We will not talk about the proof of this formula here. You can find proofs in most books about Lie theory, for example in William Fulton and Joe Harris. *Representation Theory: A First Course*. Springer, 1st corrected edition, 8 1999. ISBN 9780387974958

The main point is that the product of two group elements yields another group element. Therefore, the Baker-Campbell-Hausdorff formula tells us how we have to combine generators to get other generators which successfully yield elements of the group when put into the exponential function. And this is why we end up with a product rule like $A \circ B = AB - BA$.

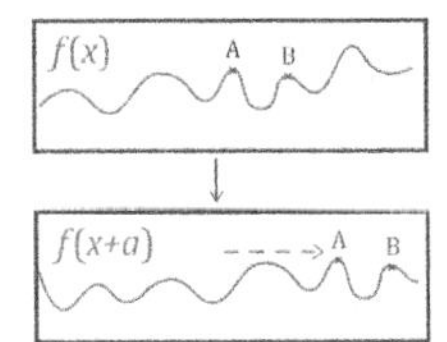

Now, what do these generators explicitly look like?

Let's consider a function $f(x,t)$ and assume that our goal is to generate a spatial translation $Tf(x,t) = f(x+a,t)$. The generator

$$G_{\text{xtrans}} = \partial_x \tag{C.9}$$

does the job:[10]

[10] Here ∂_x is a shorthand notation for the derivative $\frac{\partial}{\partial x}$.

$$\begin{aligned} e^{aG_{\text{xtrans}}} f(x,t) &= (1 + aG_{\text{xtrans}} + \frac{a^2}{2} G_{\text{xtrans}}^2 + \ldots) f(x,t) \quad \text{Eq. C.9} \\ &= (1 + a\partial_x + \frac{a^2}{2}\partial_x^2 + \ldots) f(x,t) \\ &= f(x+a,t) \end{aligned}$$

Here we used the series expansion of $e^x = \sum_{n=0}^{\infty} \frac{x^n}{n!}$ and in the last step, we used that in the second to last line we have exactly the Taylor expansion of $f(x+a,t)$.[11] Alternatively consider an infinitesimal translation: $a \to \epsilon$ with $\epsilon \ll 1$. We

[11] If you're unfamiliar with the Taylor series expansion, have a look at Appendix E.

then have

$$
\begin{aligned}
e^{\epsilon G_{\text{xtrans}}} f(x,t) &= (1 + \epsilon G_{\text{xtrans}} + \frac{\epsilon^2}{2} G_{\text{xtrans}}^2 + \ldots) f(x,t) \\
&= (1 + \epsilon G_{\text{xtrans}}) f(x,t) \\
&= (1 + \epsilon \partial_x) f(x,t) \\
&= f(x,t) + \epsilon \partial_x f(x,t) \\
&= f(x + \epsilon, t) \, .
\end{aligned}
$$

$\epsilon^2 = 0$

Eq. C.9

Here $\partial_x f(x,t)$ is the **rate of change** of $f(x,t)$ in the x-direction. In other words, this term tells us how much $f(x,t)$ changes when we move in the x-direction. If we multiply this rate of change with the distance that we move in the x-direction - here ϵ - we end up with the total change of $f(x,t)$ if we move by ϵ in the x-direction. Thus, $f(x,t) + \epsilon \partial_x f(x,t)$ really *is* the value of f at the location $x + \epsilon$.

The bottom line is: $G_{\text{xtrans}} = \partial_x$ generates *spatial* translations.

Completely analogously $G_{\text{ttrans}} = \partial_t$ generates *temporal* translations:

$$f(x,t) \to e^{a G_{\text{ttrans}}} f(x,t) = f(x, t + a) \, .$$

What we have learned here is that generators are the crucial mathematical ingredient that we need to describe continuous symmetries. We can describe *any* continuous symmetry by acting on the function in question with the corresponding generator many many times.

PS: If you're interested in how we can understand a Lie algebra as the tangent space to the identity element of the corresponding group manifold, try `http://jakobschwichtenberg.com/lie-algebra-able-describe-group/`

D

Fiber Bundles

In this appendix, as before, no attempt at mathematical rigor is made. There are already hundreds of books and lecture notes which rigorously present fiber bundles by stringing together definitions, theorems and proofs. In contrast, the sole goal of this appendix is to connect the ideas discussed in the previous chapters with the correct mathematical notions. In other words, this appendix is a bridge between our physical intuition and the mathematical literature.

Fiber bundles are interesting because they are precisely the mathematical constructions we used in the financial toy model and in quantum mechanics.

To understand fiber bundles, we need to understand how we can calculate the product of two spaces. Intuitively, by calculating the product of two spaces, we add a copy of one entire space to each point of a second space.

For example, the product of two lines yields a rectangle;

more precisely, a rectangular surface.

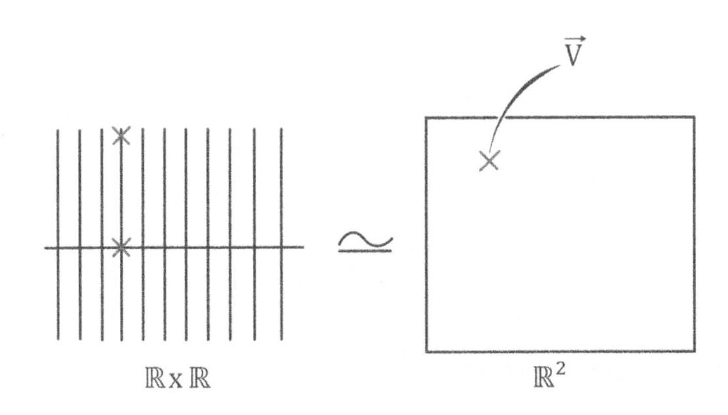

And the product of two circles yields a torus; more specifically, the surface of a torus.[1]

[1] Take note that it doesn't matter whether we imagine that we attach copies of the space A to each point of the space B or that we attach a copy of B to each point in A.

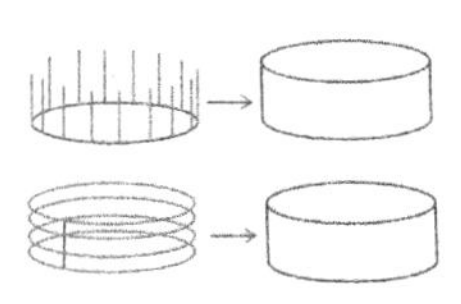

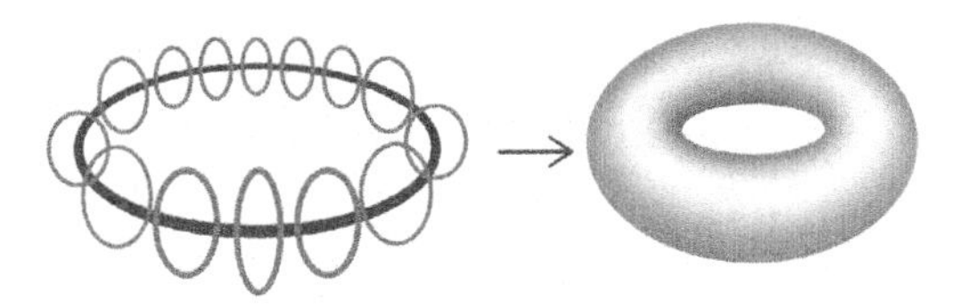

In the financial toy model, we multiply spacetime with our money space. This means we attach a copy of money space to each point of spacetime. In quantum mechanics, we multiply spacetime with charge space.[2] This means we attach a copy of charge space to each point of spacetime.

[2] We discuss spacetime in detail in Chapter 5.

Since spacetime is four-dimensional and our money space and charge space are one-dimensional, we can't really draw these products. But, if spacetime were two-dimensional, we would obtain the pictures included in the previous chapters.

Now, the whole point of this construction is that we add a copy of one space F to each point of another space B. In general, we call B the base space and F the fiber. The complete geometrical object we get by attaching a fiber F to each point of the base space B is called a fiber bundle.

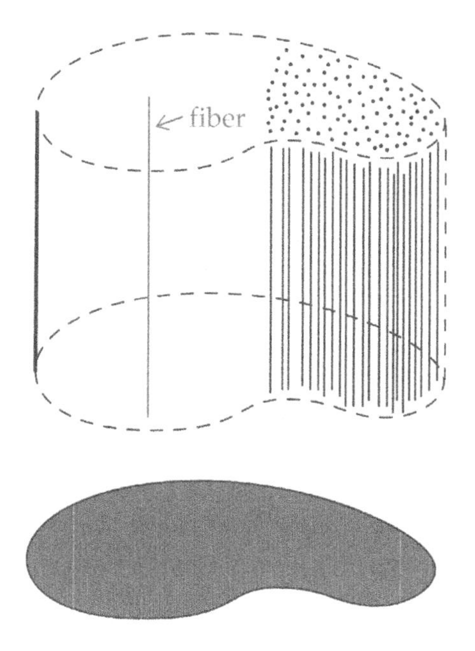

In the financial toy model and in quantum mechanics, our fibers are one-dimensional (a line and a circle respectively).[3] But we can use higher-dimensional fibers, too.[4]

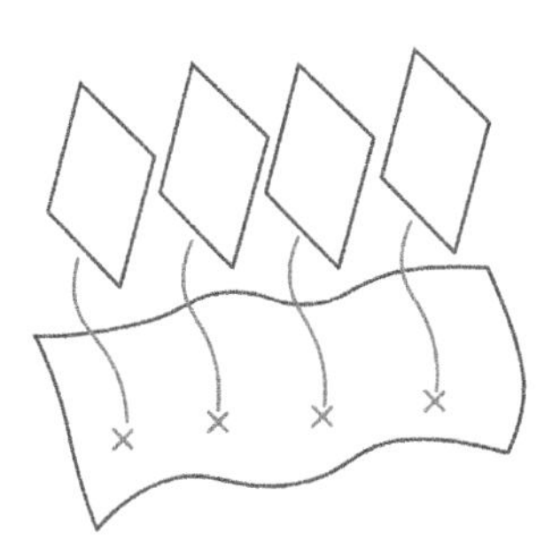

[3] The dimension of a space corresponds to the smallest number of coordinates required to localize an object in the given space. To describe the location of an object on a line, we only need one number, such as its distance along the line from the origin. Similarly, to describe an object which moves on a circle, we only need the angle φ.

[4] For isospin space and color space, our fibers are higher-dimensional.

Another important idea is that there are different kinds of fiber bundles. In particular, not every fiber bundle is a simple product of the form $B \times F$. While locally, every fiber bundle looks like a trivial product $B \times F$, there are bundles with a non-trivial global structure. The most famous example of such a non-trivial bundle is a Moebius strip. The trivial bundle we get by attaching a line to each point of a circle looks like a cylinder. But we can also imagine that there is a twist in the fibers as we move around the circle. The geometrical object we get if there is a twist in the fibers is a Moebius strip.

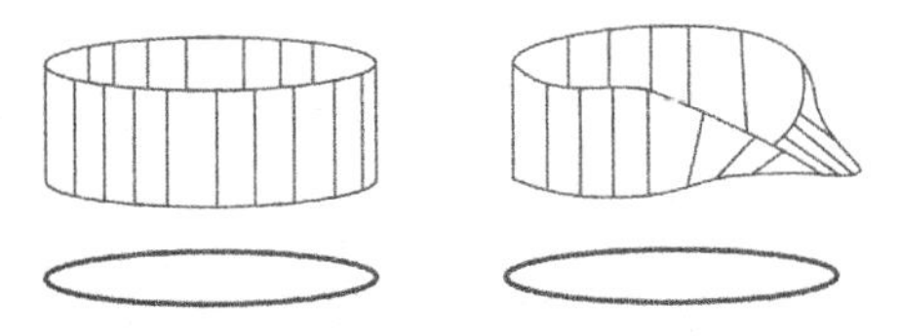

The defining feature of such a non-trivial fiber bundle is that we can't use the same coordinate system to describe locations on it globally. Instead, we need at least two coordinate systems and a rule which glues these patches together.[5]

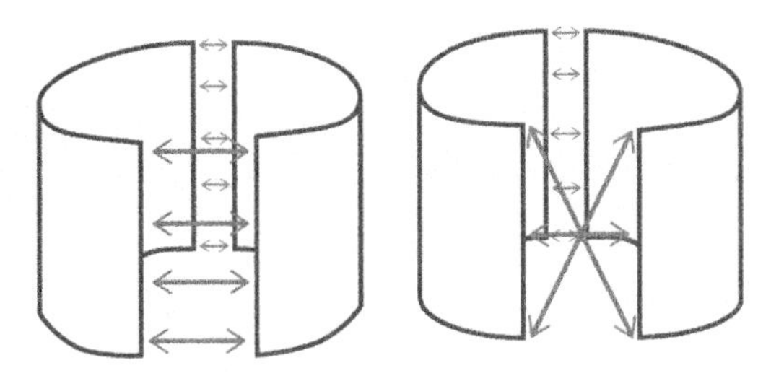

With this in mind, we can classify fiber bundles as follows:[6]

▷ **Type-1 bundles**: (topologically) trivial bundles with zero curvature.

[5] We will talk about why we are interested in such strange constructions below.

[6] We will talk about curvature below. Moreover, topology is the branch of mathematics which deals with aspects of geometrical objects that remain unchanged as we deform the object. In other words, in topology, we imagine that each object is made of some mouldable clay which we can deform however we like. (However, it's not allowed to puncture or heal nor to add or remove holes). Any two geometrical objects which can be transformed into each other this way are topologically equivalent. For example, a doughnut and a coffee mug are topologically equivalent.

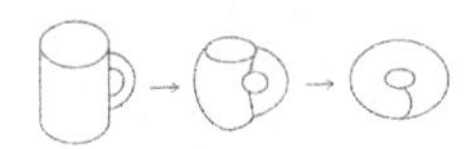

An object is topologically non-trivial whenever there is some knot that we can't get rid of by deforming the object.

▷ **Type-2 bundles**: (topologically) trivial bundles with non-zero curvature.

▷ **Type-3 bundles**: (topologically) non-trivial bundles.

Each type is important in physics and corresponds to a different situation.

▷ Type-1 bundles describe systems with a vanishing electromagnetic field strength. In the financial toy model, we have a trivial bundle with zero curvature whenever there are no arbitrage opportunities.

▷ Type-2 bundles describe systems with non-zero electromagnetic field strength. In the financial toy model, we have a trivial bundle with non-zero curvature whenever there *are* arbitrage opportunities.

▷ Type-3 bundles describe systems in which topological objects like a magnetic monopole are present.[7]

[7] A non-trivial color space bundle is necessary to describe instantons.

In this book, however, we will only talk about type-1 and type-2 bundles. To properly understand non-trivial bundles, we need quite a bit of topology and discussing this would lead us too far astray. But there are many great resources to learn more about non-trivial bundles and several are listed in Chapter 6.

Nevertheless, it is important to keep in mind that fiber bundles are a *generalization* of the direct product of spaces. In general, fiber bundles only locally look like a direct product. But globally, they can be non-trivial.

Before we move on, there is one final distinction we need to talk about. It is conventional to give special names to fiber bundles depending on what kind of space F we attach to each point of the base space B:[8]

[8] There are lots of other types of bundles. For example, the fibers of a tangent bundle are the tangent spaces T_pM at each point of the base manifold M. Tangent bundles are important in the Lagrangian formulation of classical mechanics.

▷ If F is a vector space, we call the resulting bundle a **vector bundle**.

▷ If F is a Lie group, we call the resulting bundle a **principal bundle**.[9]

[9] In the previous section we learned that we can understand Lie groups geometrically. For example, the group $U(1)$ is geometrically a circle. Therefore, we can imagine that the $U(1)$-principal bundle consists of circles which we attach to each point of the base space.

In physics, we need both. To understand why, we need to recall an extremely important fact about elementary particles.

Different elementary particles are influenced differently by the structure of the various internal spaces.[10] This is necessarily the case because otherwise we couldn't distinguish them. After all, we *define* elementary particles depending on how they are influenced by the structure of the internal spaces.

[10] For example, an electron is influenced by the structure of charge space, while a neutrino isn't.

How can we describe this using fiber bundles?

First of all, recall that in Section C.2 we discussed that we can understand symmetries geometrically. Each symmetry group corresponds to a particular geometrical object. Therefore, to construct the arena we use to describe elementary particles and their interactions, we can imagine that we attach a copy of the fundamental symmetry groups to each point in spacetime.[11]

[11] For example, $U(1)$ corresponds to a unit circle and $SU(2)$ to a sphere.

For concreteness, let's focus on $SU(2)$ for a moment.[12] To construct the arena we use to describe weak interactions, we therefore imagine that we attach a copy of $SU(2)$ to each point in spacetime.

[12] The construction for the remaining symmetry groups works analogously. We choose $SU(2)$ because it's sufficiently non-trivial (unlike $U(1)$) but not too complicated (unlike $SU(3)$).

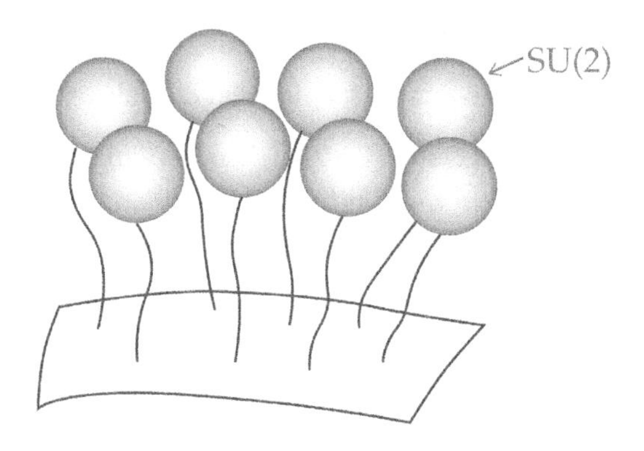

This is a principal bundle. And this principal bundle is our starting point.[13]

[13] Geometrically, $SU(2)$ is the three-sphere S^3. The three-sphere is the surface of a four-dimensional ball. We can't draw in four dimensions and therefore the two-sphere S^2 (the surface of a three-dimensional ball) is drawn here.

Now, some particles are influenced by the structure of isospin space while others aren't. Moreover, we can imagine that there are additional elementary particles which would be influenced even more strongly than the particles we know so far. This would be analogous to how a particle with a large electric charge is influenced more strongly by the structure of charge space than a particle with a small charge.

The branch of mathematics which allows us to describe this is known as **representation theory**.

D.1 Representation Theory

Representation theory is a part of group theory and the fundamental task is to study how a given group acts on different objects.[14] For beginners, this is usually an extremely confusing idea. For example:

[14] Once more, we will not discuss any details and focus on the key ideas. Group theory is discussed in Appendix C and a detailed discussion of representation theory can be found in

Jakob Schwichtenberg. *Physics from Symmetry*. Springer, Cham, Switzerland, 2018b. ISBN 978-3319666303

- ▷ Elements of $U(1)$ are defined as transformations of the form e^{ix}. So, of course, $U(1)$ acts on complex functions.
- ▷ Elements of $SU(2)$ are (2×2) matrices with unit determinant that fulfill the condition $M^\dagger M = 1$. So, of course, $SU(2)$ acts on complex column vectors with two entries.

But we can understand groups in a more abstract sense, too. In fact, we *need* this abstract perspective to make sense of modern physics. We know that not all particles are affected equally by the structure of our internal spaces. For example, an $SU(2)$ transformation rotates an electron into a neutrino and vice versa. But, there are also particles which do not get affected by the structure of isospin space.[15] Therefore, an $SU(2)$ transformation will not have any effect on them.

[15] A subtlety is that there are right-chiral partners of all fermions. Chirality is a property of elementary particles closely connected to their spin. For massless particles, chirality is equal to the projection of the particle's spin onto its momentum vector. In this context, right-chiral partners are interesting because they do not carry isospin. They therefore are not influenced by the structure of isospin space. This is the only thing we will focus on here. For example, a left-chiral electron e_L carries isospin, while a right-chiral electron e_R does not. For all fermions right-chiral partners have been observed already except for the neutrinos. (The connection between the elementary particle property of chirality (in its observed relation to spacetime) and the internal property of isospin is one of the unsolved mysteries in modern physics.)

To make sense of this, we need to forget for a moment that we define groups in the first place as a concrete set of transformations acting on specific objects. Instead, we now only remember that each group is a certain geometrical object. Each point on this geometrical object corresponds to one specific group element.

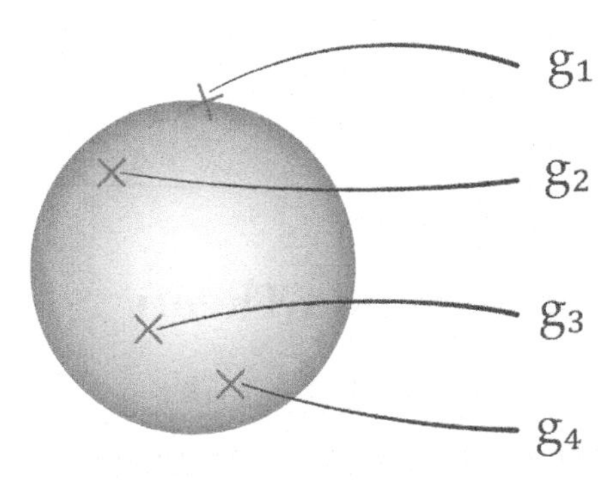

This is how we can abstractly view our group. The main idea of representation theory is that we can find maps from this abstract geometrical object to the specific linear transformations on different vector spaces.[16] In practice, this means that we try to find a way to represent our group elements as matrices which act on the elements of different vector spaces.

[16] Of course, we don't try to find just any map, but a map which fulfills specific conditions. These conditions ensure that the group properties are preserved.

A representation is a map between any group element g of a group G and a linear transformation $T(g)$ of some vector-space V in such a way that:

- $T(e) = I$ (The identity element of the group transforms nothing at all.)
- $T(g^{-1}) = (T(g))^{-1}$ (Every inverse element is mapped to the corresponding inverse transformation.)
- $T(g) \circ T(h) = T(gh)$ (The combination of transformations corresponding to g and h is the same as the transformation corresponding to the point gh)

In mathematical terms, we say that a map with these properties is a homomorphism.

For example, we can pick some vector space like $\mathbb{C}^2$ and then search for maps which yield for each point of our abstract geometrical object a concrete linear transformation which acts on the elements of $\mathbb{C}^2$. The elements of $\mathbb{C}^2$ are complex column vectors with two entries. Linear transformations acting on them are therefore (2×2) matrices. This way we get the standard representation of $SU(2)$.

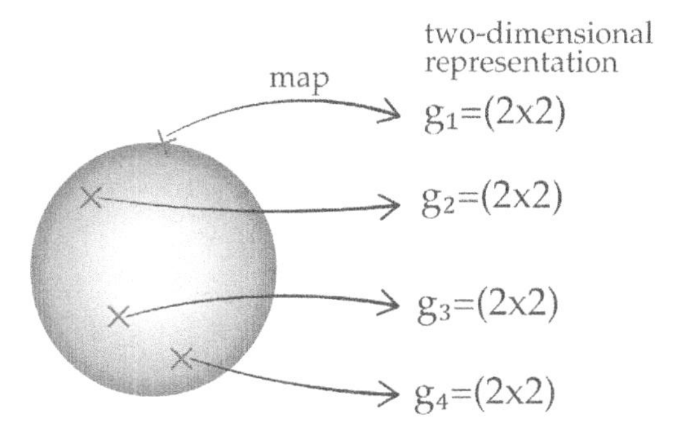

But we can also pick some other vector space like $\mathbb{C}^3$ and then search for maps onto the linear transformation on $\mathbb{C}^3$. Elements of $\mathbb{C}^3$ are complex column vectors with three entries. Therefore linear transformations acting on them are (3×3) matrices.

In other words, each element of $SU(2)$ gets mapped to one specific (3×3) matrix. This yields the three-dimensional representation of $SU(2)$![17]

[17] It is crucial to understand that this three-dimensional representation of $SU(2)$ is not simply $SU(3)$. The three-dimensional representation of $SU(3)$ looks very different from the three-dimensional representation of $SU(2)$.

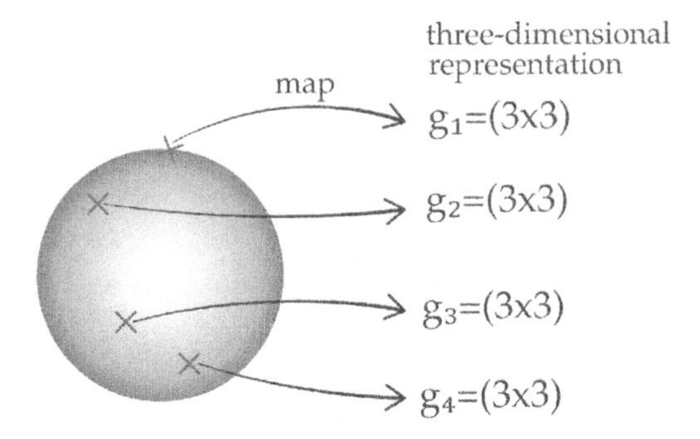

We can also pick an extremely simple vector space like $\mathbb{C}$. But the only way we can represent $SU(2)$ as linear operators acting on elements of $\mathbb{C}$ is to map them all to the identity which changes nothing at all. This means that the one-dimensional representation of $SU(2)$ is trivial and consists only of the element 1 which leaves elements of $\mathbb{C}$ completely unchanged.

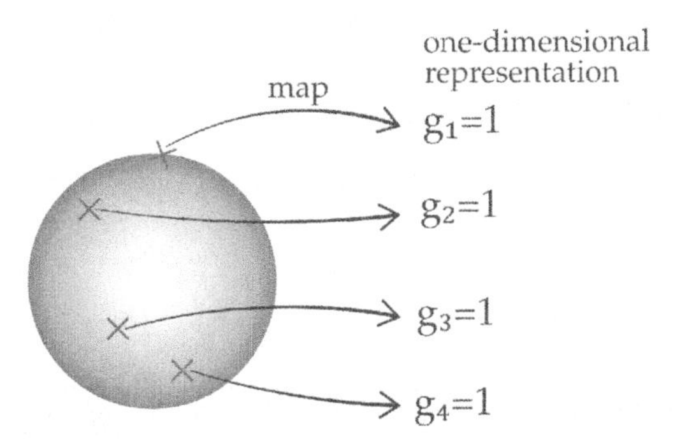

This is exactly what we need when we try to describe elementary particles which do not get influenced by $SU(2)$ at all. When $SU(2)$ acts on the objects that we use to describe elementary particles without isospin, we need to represent its elements using the one-dimensional representation. In contrast, when $SU(2)$ acts on objects we use to describe particles carrying isospin, we need to use a non-trivial representation. It is conventional to say that particles without isospin *live* in the one-dimensional representation of $SU(2)$, while particles like the (left-chiral) electron and (left-chiral) electron-neutrino *live* in the two-dimensional representation.[18]

[18] Take note that for a particle with electric charge q, elements of the corresponding $U(1)$ representation look like $e^{iq\epsilon}$. Therefore, the $U(1)$ representation for particles without electric charge $q = 0$ is simply $e^{i0\epsilon} = 1$. In other words, the $q = 0$ representation of $U(1)$ is trivial, analogous to the trivial representation of $SU(2)$ and is therefore exactly what we need to describe particles that are not affected by the structure of charge space.

Statements like this often sound extremely abstract. It is therefore important to keep in mind that the main idea is really that different elementary particles experience the structure of our internal spaces differently. Representation theory is the mathematical toolkit we can use to take this fact into account.

With this in mind, let's get back to our fiber bundles.

D.2 Associated Bundles

Before we started talking about representation theory, we noted that the object we start with in physics is called a principal bundle. A principal bundle consists of a copy of a specific group G attached to each point in spacetime.[19]

[19] Since in fundamental physics there are multiple internal symmetries, we attach all these groups to each point in spacetime.

Now, we somehow need to take the lessons learned in the previous section into account.

We can do this by using so-called **associated vector bundles**. The idea here is that a principal bundle has an associated vector bundle for *each* of its representations.

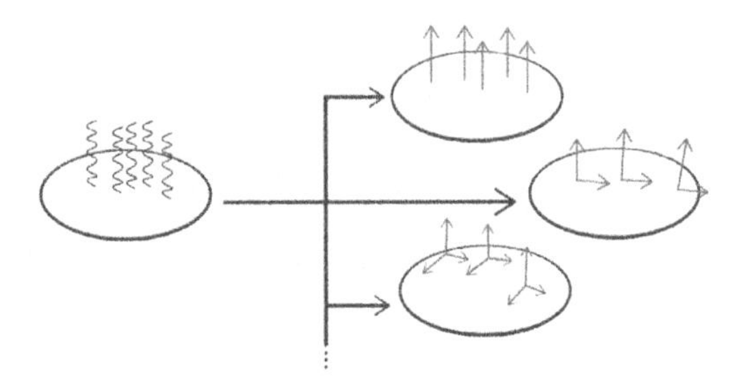

For the one-dimensional representation, we add a copy of the corresponding one-dimensional vector space that the group elements act on to each point of the base space. Similarly, for the two-dimensional representation, we add a copy of the corresponding two-dimensional vector space and for the three-dimensional representation, we add a copy of the three-dimensional vector space.

Intuitively, these associated vector spaces represent how the internal space (e.g. isospin space) looks like for a particular particle. A particle without isospin, only sees the one-dimensional version[20]

[20] An example of a particle without isospin is the right-chiral electron.

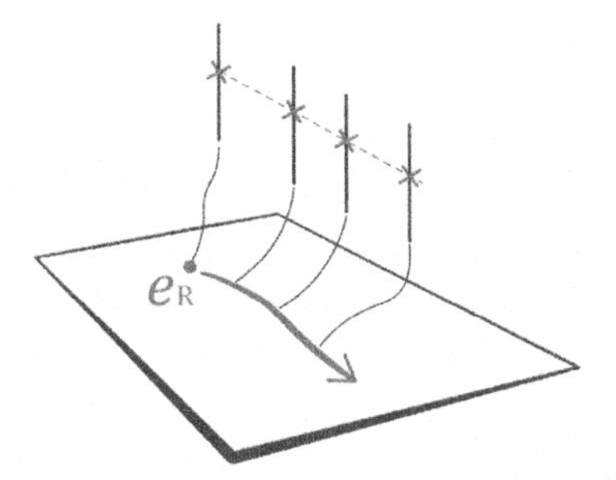

In contrast, the left-chiral electron/electron-neutrino state sees the two-dimensional version:

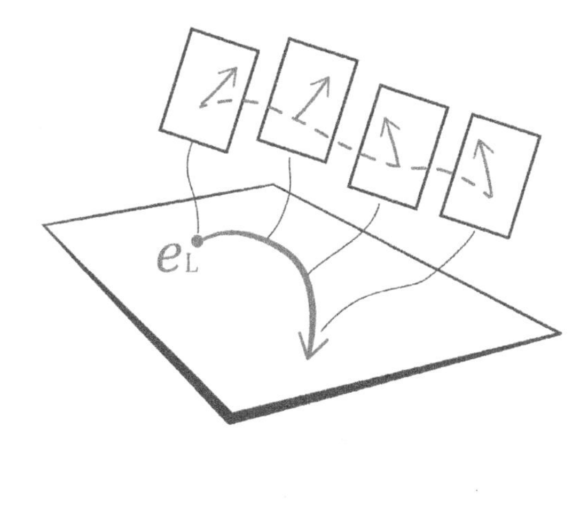

Now that we have a rough understanding of the arena we use in physics, it's time to talk about the objects which glue our fibers together. These objects tell us how a given particle moves from fiber to fiber as it moves through spacetime. In other words, they tell us which path in a fiber a given particle follows as it follows a specific trajectory in spacetime.

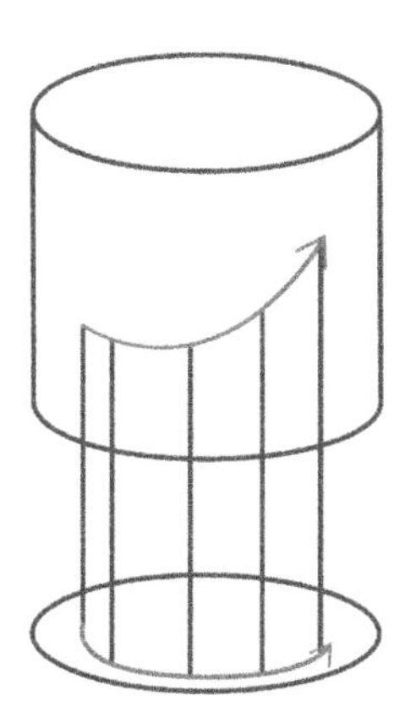

Mathematically, we call **connections** A_μ, the objects which tell us how particles move from fiber to fiber.[21] And tech-

[21] In mathematical terms, the connections live in the adjoint representation of the corresponding symmetry group. This is an important representation for each Lie group because it tells us how the group acts on its own Lie algebra. While a Lie group is, in general, a complicated manifold, the corresponding Lie algebra is always a vector space. Therefore, we can investigate how any given group acts on this particular vector space.

nically, a connection allows us to parallel transport a given object. Therefore, before we can talk about connections, we need to talk about parallel transport.

D.3 Parallel Transport

A crucial task whenever we are dealing with a non-trivial geometrical arena is to understand how we can move a given object around *consistently*. We need a method that allows us to compare two objects at different locations.[22] This is not an easy task because any apparent difference between two objects could simply be the result of different local conventions or a result of the non-trivial structure of our geometrical arena. This means that although two objects may look very different at first glance, we may find that they are equivalent when we bring them to the same location.

[22] Reminder: We compare objects at different locations whenever we calculate the derivative.

For example, 100 units of the local currency in some country A and 1000 units of the local currency in another may look very different at first glance. But if the exchange rate is actually $10 : 1$ these two amounts of money are actually equivalent. The exchange rate is an example of a connection which is the topic of the next section. In this section, however, we want to understand a little better the process of moving objects around consistently.

Imagine that you're walking on a curved surface while holding a stick in your hand. Your task is to bring the stick to another location while keeping its orientation unchanged (at every step) relative to the local surface. When you successfully do this, you parallel transport the stick.

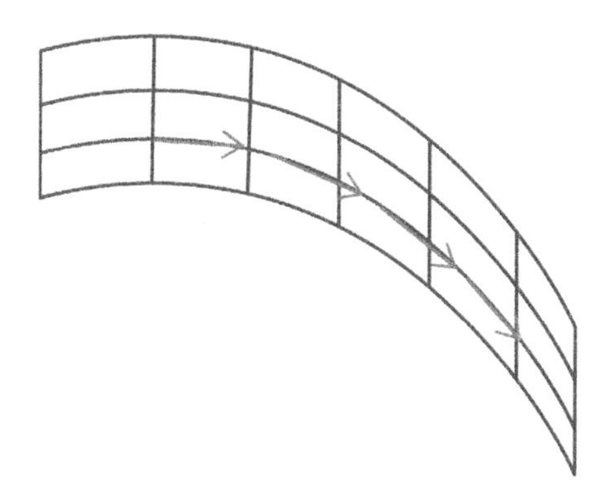

In contrast, a non-parallel transport of the stick looks as follows

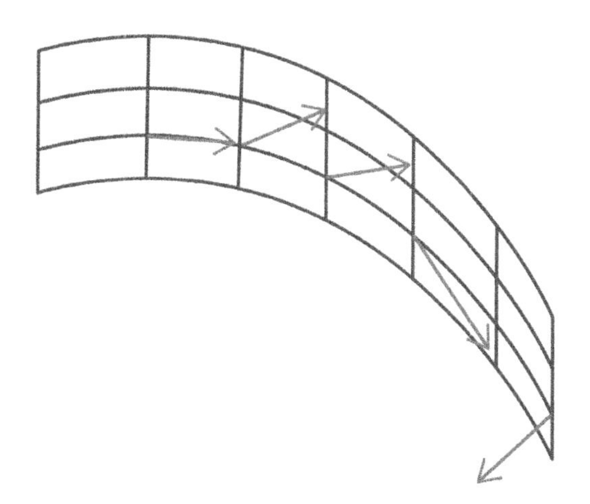

Moreover, take note that to understand parallel transport you really need to imagine that you're really walking along the surface. The transport shown in the following picture is not parallel even though it looks parallel from the outside:

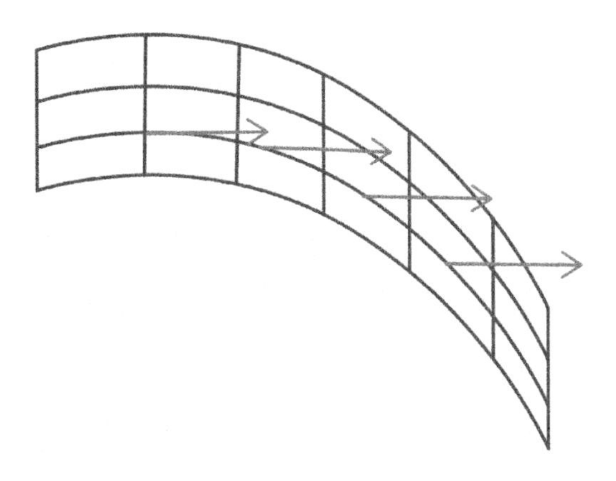

With this intuitive idea in mind, let's talk about connections.

D.4 Connections

In the financial toy model, after the introduction of the bookkeepers A_μ, we can perform local dilations[23]

[23] To unclutter the notation, we restrict ourselves to one spatial dimension here.

$$f(x) = e^{\epsilon(x)} \tag{D.1}$$

since our bookkeepers adjust accordingly (Eq. 2.21). Analogously, in quantum mechanics we can then perform local phase shifts

$$f(x) = e^{i\epsilon(x)} \tag{D.2}$$

as soon as we introduce the bookkeepers A_μ.

The crucial point is that now our transformation parameters ϵ are functions of the location x.

In other words, we can shift the prices or analogously the phase at each point x in space by a different amount. Without the bookkeepers, only global shifts are permitted which shift the prices, or analogously the phases, everywhere by exactly the same amount.

We then exploited this knowledge to understand the internal space of the toy model and of quantum mechanics. In the toy model, our internal space at each location is the real line $\mathbb{R}^+$. This line represents all possible prices.

In quantum mechanics, the internal space at each location is a unit circle in the complex plane. This circle represents all possible phase factors. The total internal space is all these individual internal spaces taken together.

A crucial ingredient to make this possible are the bookkeepers A_μ which glue the individual spaces together.

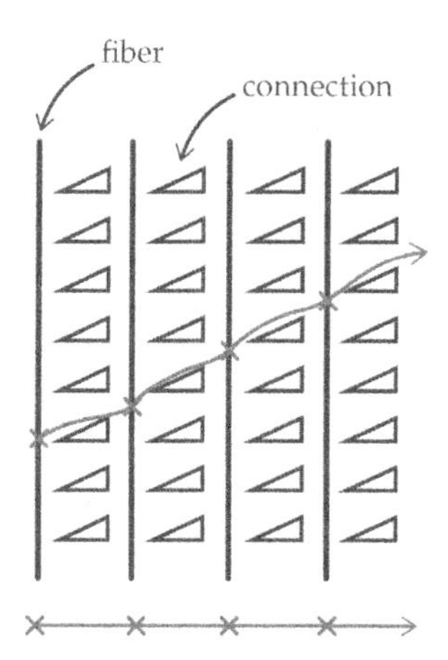

Mathematically, the bookkeepers A_μ are called **connections**. This name makes sense because the connections are what we need to consistently describe how goods (like copper) and elementary particles move around.

This is non-trivial, as we have discussed extensively in the context of the financial toy model, because there can be local currencies. This means that it is not immediately clear what a given amount of money is worth in terms of the currency of a neighboring country. In financial terms, a connection tells us the exchange rates between currencies. Formulated differently, we need the connections to keep track of local shifts. Nothing physical in the system can be influenced by local phase or price shifts because these quantities do not have any meaning in an absolute sense. Every time such a shift happens, the bookkeepers adjust accordingly such that the dynamics stay the same.

But connections are not only important to allow arbitrary local transformations. Connections are absolutely essential whenever there is curvature. In the context of the financial toy model, we called curvature a currency arbitrage opportunity $F_{\mu\nu}(t, \vec{x})$. In electrodynamics, we call curvature the electromagnetic field $F_{\mu\nu}(t, \vec{x})$. We will talk about how the name curvature comes about below, but first let's clarify this important point.

In the financial toy model, an arbitrage opportunity has, so to speak, an absolute reality, it is something physical which directly influences the dynamics; it is not something which relies on local conventions. Hence, it cannot go away through adjustments of the local currencies.

Formulated differently, when there is an arbitrage opportunity ($F_{\mu\nu} \neq 0$), it is impossible to choose a unique (global) currency. If it were possible to choose a global currency, all exchange rates would be trivial ($A_\mu = 0$ everywhere) and there would therefore be no arbitrage opportunities. We can see this immediately by recalling the relationship between the exchange rates and the gain factors associated with currency arbitrage opportunities (Eq. 2.22):

$$F_{\mu\nu}(t,\vec{x}) \equiv \frac{\partial A_\nu(t,\vec{x})}{\partial x^\mu} - \frac{\partial A_\mu(t,\vec{x})}{\partial x^\nu}. \tag{D.3}$$

The existence of an arbitrage opportunity means $F_{\mu\nu} \neq 0$ somewhere, and therefore A_μ has to be non-vanishing.

In other words, since arbitrage opportunities cannot go away through changes of local conventions, we can conclude that whenever there is one, there will be local currencies and non-trivial exchange rates. A non-trivial exchange rate means that our description of the dynamics *must* include the bookkeepers A_μ.

We can also imagine that countries use local currencies but there is no currency arbitrage opportunity. In this case, the bookkeepers are not essential parts of the model because we are free to use the symmetry to introduce a global currency. In other words, by making use of local gauge transformations, we can make the exchange rates trivial everywhere and hence there is no need for us to use the bookkeepers A_μ at all. Moreover, we learned that as soon as there is copper located *somewhere* in the system, there will be an arbitrage opportunity and we therefore *need* the bookkeepers A_μ.[24]

[24] This is what Eq. 2.34 tells us.

We have exactly the same situation in quantum mechanics. On the one hand, we need the bookkeepers to allow arbitrary local phase shifts. But these local phase shifts do not influence the dynamics of the system and we could easily live without them.[25]

[25] Formulated differently, the freedom to perform local phase shifts and to introduce local currencies is nice to have but not essential.

But on the other hand, as soon as there is a charged particle, there is a non-zero electromagnetic field $F_{\mu\nu} \neq 0$ and the bookkeepers A_μ are essential.[26] Formulated more technically, if there is no charged object present in the system, we can use local gauge transformations to get rid of the

[26] Again, this is what Eq. 2.34 and Eq. 2.22 tell us.

bookkeepers A_μ everywhere. But the presence of just one charged object in the system is enough to make bookkeepers essential parts of the system and we can't get rid of them everywhere at the same time.

Using the terminology introduced above, we can say that connections are only essential for type-2 bundles (and type-3 bundles). For type-1 bundles, there is no curvature and therefore we can use our local symmetry to set the connections to zero everywhere.

D.5 Curvature

Now, why exactly do we call the quantity

$$F_{\mu\nu}(t,\vec{x}) \equiv \frac{\partial A_\nu(t,\vec{x})}{\partial x^\mu} - \frac{\partial A_\mu(t,\vec{x})}{\partial x^\nu}. \tag{D.4}$$

the curvature?

To understand this, we need to discuss how we can measure the curvature of a given space in general. While the details always depend on the space at hand, the main idea is always that we move an object along a closed loop and then compare the state of the object after the journey with its state at the beginning.

The prototypical example of a curved space is a sphere. To understand how we can measure the curvature of a sphere imagine that you are walking on the sphere while holding a stick in your hand. While you walk, you always do your best to keep the stick straight.[27] An important observation is that if the space you are moving in is curved, it's possible that the stick, even though it returns to its starting location, does not end in its starting orientation if you move along a closed loop.

[27] If you do this you're parallel transporting the stick. In our framework, the connections A_μ make sure that we are transporting our objects correctly. This is crucial because we want to measure a real feature of the system (the curvature) and want to avoid that any local convention distorts our result.

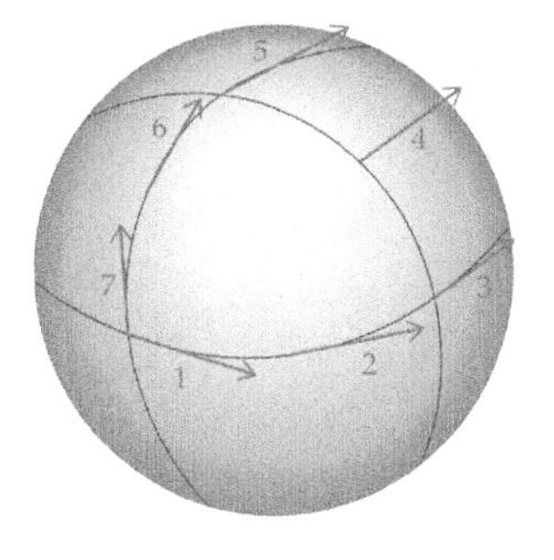

Hence, the difference between the original vector and the

vector which was parallel transported along a closed loop encodes information about the curvature. In contrast, if we parallel transport a vector in a flat space, its initial and final state will be equivalent:

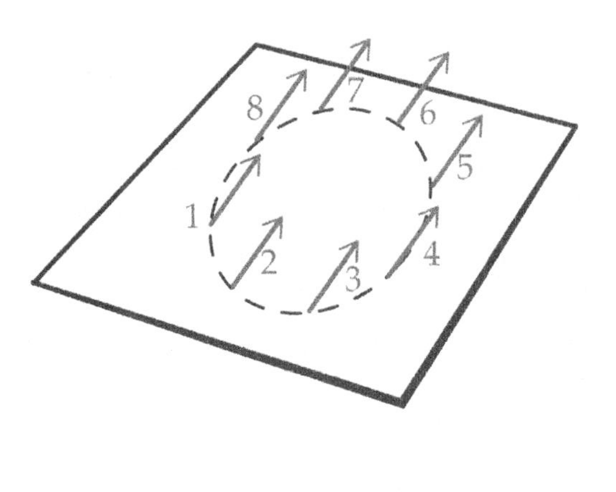

Completely analogously, if there is an arbitrage opportunity in the financial toy model, we can move a specific amount of money along a closed loop and end up with more money than we started with.

This indicates that we are dealing with a curved internal space. This is precisely how we defined the gain factor in Eq. 2.9.[28]

[28] Reminder: Eq. 2.9 reads

$$F_{ij}(\vec{n}) = A_j(\vec{n}+\vec{e}_i) - A_j(\vec{n}) - [A_i(\vec{n}+\vec{e}_j) - A_i(\vec{n})] .$$

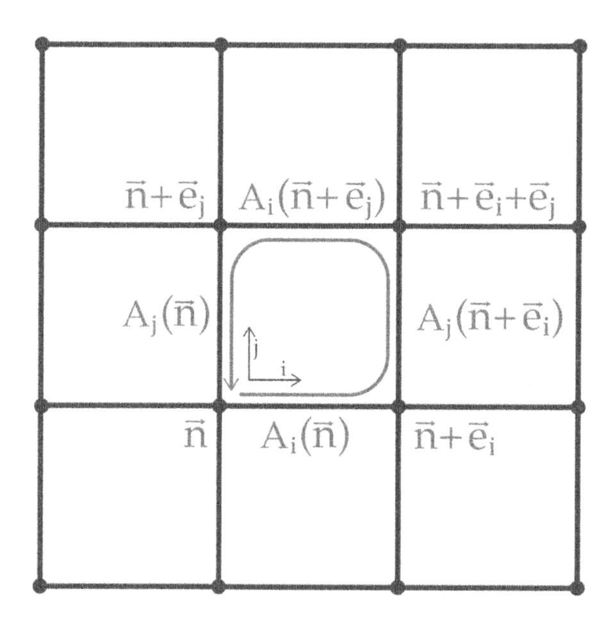

In this sense, this gain factor is a measure of the curvature of the internal money space. And, in exactly the same sense, we can say that the electromagnetic field strength tensor $F_{\mu\nu}(t, \vec{x})$ encodes how much charge space is curved.

This is one of the main observations at the heart of modern physics.

E

Taylor Expansion

The Taylor expansion is one of the most useful mathematical tools and we need it in physics all the time to simplify complicated systems and equations.

We can understand the basic idea as follows:

Imagine you sit in your car and wonder what your exact location $l(t)$ will be in 10 minutes: $l(t_0 + 10 \text{ minutes}) = ?$

▷ A first estimate is that your location will be exactly your *current* location

$$l(t_0 + 10 \text{ minutes}) \approx l(t_0) .$$

Given how large the universe is and thus how many possible locations there are, this is certainly not too bad.

▷ If you want to do a bit better than that, you can also include your *current* velocity $\dot{l}(t_0) \equiv \partial_t l(t)\big|_{t_0}$.[1] The total distance you will travel in 10 minutes if you continue to

[1] Here ∂_t is a shorthand notation for $\frac{\partial}{\partial t}$ and $\partial_t l(t)$ yields the velocity (rate of change). After taking the derivative, we evaluate the velocity function $\dot{l}(t) \equiv \partial_t l(t)$ at t_0: $\dot{l}(t_0) = \partial_t l(t)\big|_{t_0}$.

move at your current velocity is this velocity times 10 minutes: $\dot{l}(t_0) \times 10$ minutes. Therefore, your second estimate is your current location plus the velocity you are traveling times 10 minutes

$$l(t_0 + 10 \text{ minutes}) \approx l(t_0) + \dot{l}(t_0) \times 10 \text{ minutes}. \quad \text{(E.1)}$$

▷ If you want to get an even better estimate you need to take into account that your velocity can possibly change. The rate of change of the velocity $\ddot{l}(t_0) = \partial_t^2 l(t)\big|_{t_0}$ is what we call acceleration. So in this third step you additionally take your *current* acceleration into account[2]

$$l(t_0 + 10 \text{ minutes}) \approx l(t_0) + \dot{l}(t_0) \times 10 \text{ minutes} + \frac{1}{2}\ddot{l}(t_0) \times (10 \text{ minutes})^2.$$

[2] The factor $\frac{1}{2}$ and that we need to square the 10 minutes follows since, to get from an acceleration to a location, we have to integrate twice:

$$\int dt \int dt \ddot{x}(t_0) = \int dt \ddot{x}(t_0) t = \frac{1}{2}\ddot{x}(t_0)t^2$$

where $\ddot{x}(t_0)$ is the value of the acceleration at $t = t_0$ (= a constant).

▷ Our estimate will still not yield the perfect final location since, additionally, we need to take into account that our acceleration could change during the 10 minutes. We could therefore additionally take the current rate of change of our acceleration into account.

This game never ends and the only limiting factor is how precisely we want to estimate our future location. For many real-world purposes, our first order approximation (Eq. E.1) would already be perfectly sufficient.

The procedure described above is exactly the motivation behind the Taylor expansion. In general, we want to estimate the value of some function $f(x)$ at some value of x by using our knowledge of the function's value at some fixed point a. The **Taylor series** then reads[3]

[3] Here the superscript (n), within brackets, denotes the n-th derivative. For example $f^{(0)} = f$ and $f^{(1)}$ is $\partial_x f$.

$$f(x) = \sum_{n=0}^{\infty} \frac{f^{(n)}(a)(x-a)^n}{n!}$$
$$= \frac{f^{(0)}(a)(x-a)^0}{0!} + \frac{f^{(1)}(a)(x-a)^1}{1!} + \frac{f^{(2)}(a)(x-a)^2}{2!}$$
$$+ \frac{f^{(3)}(a)(x-a)^3}{3!} + \dots, \tag{E.2}$$

where $f(a)$ is the value of the function at the point a we are expanding around. Moreover, $x - a$ is analogous to the 10 minute timespan we considered above. If we want to know the location at $x =$ 5:10 pm by using our knowledge at $a =$ 5:00 pm, we get $x - a =$ 5:10 pm $-$ 5:00 pm $=$ 10 minutes. Therefore, this equation is completely analogous to our estimate of the future location we considered previously.

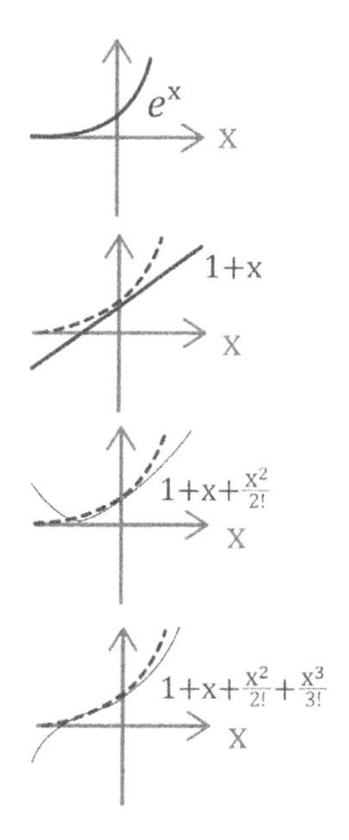

To understand the Taylor expansion a bit better, it is helpful to look at concrete examples.

We start with one of the simplest but most important examples: the exponential function. Putting $f(x) = e^x$ into Eq. E.2 yields[4]

$$e^x = \sum_{n=0}^{\infty} \frac{(e^x)^{(n)}\Big|_{x=0}(x-0)^n}{n!}$$

[4] The notation $(e^x)^{(n)}\Big|_{x=0}$ means that we take the n-th derivative of the function e^x and then plug in $x = 0$ into the result.

The crucial puzzle pieces that we need are $(\mathrm{e}^x)' = \mathrm{e}^x$ and $\mathrm{e}^0 = 1$. Using these relations and substituting into the previous equation yields

$$\mathrm{e}^x = \sum_{n=0}^{\infty} \frac{\mathrm{e}^0(x-0)^n}{n!} = \sum_{n=0}^{\infty} \frac{x^n}{n!} \tag{E.3}$$

This result can be used as a definition of e^x.

Next, let's assume that the function we want to approximate is $\sin(x)$ and we want to expand it around $x = 0$.

Putting $f(x) = \sin(x)$ into Eq. E.2 yields[5]

$$\sin(x) = \sum_{n=0}^{\infty} \frac{\left(\sin(x)\right)^{(n)}\Big|_{x=0}(x-0)^n}{n!}$$

[5] As before, the notation $\left(\sin(x)\right)^{(n)}\Big|_{x=0}$ means that we take the n-th derivative of the function $\sin(x)$ and then plug in $x = 0$ into the result. So for example, since

$$\left(\sin(x)\right)^{(1)} = \cos(x),$$

we have,

$$\left(\sin(x)\right)^{(1)}\Big|_{x=0} = \cos(0) = 1.$$

The crucial information we therefore need is $(\sin(x))' = \cos(x)$, $(\cos(x))' = -\sin(x)$, $\cos(0) = 1$ and $\sin(0) = 0$. Because $\sin(0) = 0$, every term with even n vanishes, which we can use if we split the sum. Observe that

$$\sum_{n=0}^{\infty} n = \sum_{n=0}^{\infty}(2n+1) + \sum_{n=0}^{\infty}(2n)$$

$$1+2+3+4+5+6\ldots = 1+3+5+\ldots \quad +2+4+6+\ldots$$

Therefore, splitting the sum into even and odd terms yields

$$\begin{aligned}
\sin(x) &= \sum_{n=0}^{\infty} \frac{\left(\sin(x)\right)^{(2n+1)}\Big|_{x=0}(x-0)^{2n+1}}{(2n+1)!} \\
&\quad + \sum_{n=0}^{\infty} \frac{\left(\sin(x)\right)^{(2n)}\Big|_{x=0}(x-0)^{2n}}{(2n)!} \qquad \text{↷ } \sin(0) = 0 \\
&= \sum_{n=0}^{\infty} \frac{\left(\sin(x)\right)^{(2n+1)}\Big|_{x=0}(x-0)^{2n+1}}{(2n+1)!}. \qquad \text{(E.4)}
\end{aligned}$$

Moreover, every even derivative of $\sin(x)$ (i.e., $\sin^{(2n)}$) is again $\sin(x)$ or $-\sin(x)$. Therefore the second term vanishes since $\sin(0) = 0$. The remaining terms are odd derivatives of $\sin(x)$, which are all proportional to $\cos(x)$. We now use

$$\begin{aligned}
\sin(x)^{(1)} &= \cos(x) \\
\sin(x)^{(2)} &= \cos'(x) = -\sin(x) \\
\sin(x)^{(3)} &= -\sin'(x) = -\cos(x) \\
\sin(x)^{(4)} &= -\cos'(x) = \sin(x) \\
\sin(x)^{(5)} &= \sin'(x) = \cos(x)
\end{aligned}$$

The general pattern is $\sin^{(2n+1)}(x) = (-1)^n \cos(x)$, as you can check by putting some integer values for n into the formula[6].

[6] $\sin^{(1)}(x) = \sin^{(2\cdot 0+1)}(x) = (-1)^{\cdot 0}\cos(x) = \cos(x)$, $\sin^{(3)}(x) = \sin^{(2\cdot 1+1)}(x) = (-1)^1 \cos(x) = -\cos(x)$

Thus, we can rewrite Eq. E.4 as

$$\begin{aligned}
\sin(x) &= \sum_{n=0}^{\infty} \frac{\Big(\sin(x)\Big)^{(2n+1)}\Big|_{x=0}(x-0)^{2n+1}}{(2n+1)!} \\
&= \sum_{n=0}^{\infty} \frac{(-1)^n \cos(0)(x-0)^{2n+1}}{(2n+1)!} \qquad \text{↷ } \cos(0) = 1 \\
&= \sum_{n=0}^{\infty} \frac{(-1)^n (x)^{2n+1}}{(2n+1)!} \qquad \text{(E.5)}
\end{aligned}$$

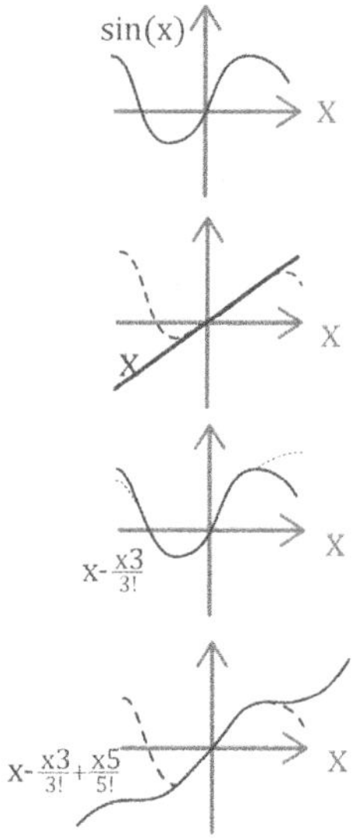

This is the Taylor expansion of $\sin(x)$, which we can also use as a definition of the sine function.

Bibliography

John Baez and Javier P. Muniain. *Gauge fields, knots, and gravity*. World Scientific, Singapore River Edge, N.J, 1994. ISBN 9789810217297.

H. J. Bernstein and A. V. Phillips. Fiber Bundles and Quantum Theory. *Sci. Am.*, 245:94–109, 1981.

Subrahmanyan Chandrasekhar. *A scientific autobiography*. World Scientific, Singapore Hackensack, NJ, 2010. ISBN 9789814299572.

Ta-Pei Cheng. *Relativity, Gravitation and Cosmology: A Basic Introduction*. Oxford University Press, 2nd edition, 1 2010. ISBN 9780199573646.

Jennifer Coopersmith. *The lazy universe*. Oxford University Press, 2017. ISBN 9780198743040.

Richard Feynman. *The Feynman lectures on physics*. Addison-Wesley, San Francisco, Calif. Harlow, 2011. ISBN 9780805390650.

Daniel Fleisch. *A student's guide to Maxwell's equations*. Cambridge University Press, 2008. ISBN 978-0521701471.

William Fulton and Joe Harris. *Representation Theory: A First Course*. Springer, 1st corrected edition, 8 1999. ISBN 9780387974958.

David Griffiths. *Introduction to quantum mechanics*. Pearson Prentice Hall, 2005. ISBN 9780131118928.

David Griffiths. *Introduction to electrodynamics*. Pearson Education Limited, Harlow, 2014. ISBN 9781292021423.

Chan Hong-Mo and Tsou Sheung Tsun. *Some elementary gauge theory concepts*. World Scientific, Singapore River Edge, N.J, 1993. ISBN 9789810210809.

Kirill Ilinski. Physics of Finance. 1997.

Kirill Ilinski. *Physics of finance: gauge modelling in non-equilibrium pricing*. John Wiley, Chichester England New York, 2001. ISBN 9780471877387.

Robert D. Klauber. *Student Friendly Quantum Field Theory*. Sandtrove Press, 2nd edition, 12 2013. ISBN 9780984513956.

Pia Malaney. The index number problem: A differential geometric approach. 1996.

Juan Maldacena. The symmetry and simplicity of the laws of physics and the Higgs boson. *Eur. J. Phys.*, 37(1):015802, 2016. DOI: 10.1088/0143-0807/37/1/015802.

David Morin. *Introduction to classical mechanics: with problems and solutions*. Cambridge University Press, 2008. ISBN 9780511808951.

Gregory Naber. *Topology, geometry and Gauge fields: foundations*. Springer Science+Business Media, LLC, New York, 2011. ISBN 9781441972538.

Roger Penrose. *The road to reality: a complete guide to the laws of the universe*. Vintage Books, New York, 2007. ISBN 9780679776314.

Jakob Schwichtenberg. *No-Nonsense Electrodynamics*. No-Nonsense Books, 2018a. ISBN 978-1790842117.

Jakob Schwichtenberg. *Physics from Symmetry*. Springer, Cham, Switzerland, 2018b. ISBN 978-3319666303.

Jakob Schwichtenberg. *No-Nonsense Quantum Mechanics*. No-Nonsense Books, 2018c. ISBN 978-1719838719.

Jakob Schwichtenberg. *No-Nonsense Classical Mechanics*. No-Nonsense Books, 2019. ISBN 978-1096195382.

John Stillwell. *Naive Lie Theory*. Springer, 1st edition, August 2008. ISBN 978-0387782140.

Samuel E. Vazquez and Simone Farinelli. Gauge invariance, geometry and arbitrage, 2009.

K. Young. Foreign exchange market as a lattice gauge theory. *American Journal of Physics*, 67(10):862–868, 1999. DOI: 10.1119/1.19139. URL https://doi.org/10.1119/1.19139.

Anthony Zee. *Einstein Gravity in a Nutshell*. Princeton University Press, 1st edition, 5 2013. ISBN 9780691145587.

Index

Made in the USA
Coppell, TX
27 July 2020

31876666R00115